Workbook

Progress in Mathematics

SADLIER-OXFORD

Rose Anita McDonnell

Catherine D. LeTourneau

Anne Veronica Burrows

Judith Ann Geschke

Francis H. Murphy

M. Winifred Kelly

with

Dr. Elinor R. Ford

Sadlier-Oxford

A Division of William H. Sadlier, Inc.

Table of Contents

Textbook Chapter	Workbook Page

10 DECIMALS: ADDITION AND SUBTRACTION

11 DECIMALS: MULTIPLICATION AND DIVISION

12 METRIC MEASUREMENT, AREA, AND VOLUME

Textbook Chapter	Workbook Page

13 RATIO, PROPORTION, AND PERCENT

14 MOVING ON: ALGEBRA

Problem-Solving Strategy: Logical Reasoning

Name ______________________

Date ______________________

Dana, Kim, and Lucinda come to school different ways. One walks, one rides her bicycle, and one rides a school bus. Dana does not ride. Lucinda lives too far to ride a bicycle to school. Who rides a bicycle to school?

	Walk	Bike	Bus
Dana	*yes*	*no*	*no*
Kim	*no*	*yes*	*no*
Lucinda	*no*	*no*	*yes*

Make a table and use logical reasoning to complete it.

Kim rides a bicycle to school.

Solve. Do your work on a separate sheet of paper.

1. Peter, Sally, and Gina order sandwiches for lunch. They order a chicken salad, a tuna fish, and a cheese sandwich. Gina does not like tuna fish. Sally does not like chicken salad or tuna fish. What sandwich does each person order for lunch?

   ____

2. Chan, Bruce, and Armando ride their bicycles to school. One bicycle is red, one is blue, and one is green. Chan does not ride a red bicycle, and Armando does not ride a red or a blue bicycle. What color bicycle does each person ride?

   ____

3. Zena buys 8 pieces of fruit. She buys fewer cantaloupes than apples and more apples than pears. She buys 3 pears. How many cantaloupes and how many apples does she buy?

   ____

4. Nick, Nora, and Nancy have pets. One has a kitten, one has a puppy, and one has a gerbil. Nora has the gerbil. Nancy is a cousin of the person with the kitten. Who has the puppy?

   ____

5. Morris, Mike, and Mickey each play one sport. One plays football, one plays baseball, and one plays basketball. Morris and the basketball player are brothers. The football player is not related to the others. Mickey's sport is played outdoors. What sport does each person play?

   ____

6. Mary, Mavis, Marsha, and Mariann are each holding a card with one of the numbers 4, 5, 6, and 7 on it. The number on each girl's card does not match the number of letters in her name. Mary is holding a card with an even number. Mavis is not holding the card with the greatest number. What number is on each girl's card?

   ____

7. Robert has 5 coins that are worth 37¢ in all. Only 1 is a quarter. What are the other coins?

   ____

8. Jennifer was born in the month whose name has the fewest letters. The date is an odd number. The sum of the digits is 2. When is Jennifer's birthday?

   ____

 Use with page 24.

Problem-Solving Strategy: Interpret the Remainder

Name ______________________

Date ______________________

The Travel Club is planning an automobile tour along the coast of California. Thirty-two people signed up for the tour. If 5 people ride in each car, how many cars will be needed for the tour?

```
    6  R2
5 ) 3 2
  - 3 0
      2
```

One car is needed for each group of 5.
There are 6 groups of 5.
But there are 2 more people.
So another car is needed.
Seven cars will be needed for the tour.

Solve. Do your work on a separate sheet of paper.

1. No more than 8 people can be assigned to each tour guide. What is the least number of tour guides needed for 55 people.

____

2. Alicia has 25 days she can use for vacation. She wants to take some 7-day trips. How many 7-day trips can Alicia take?

____

3. A tour bus is carrying 46 people. The bus stops at a restaurant where each table seats 10 people. What is the least number of tables needed to seat all the people.

____

4. Ted earns 1 vacation day for every 2 weeks he works. He has worked 15 weeks so far. How many vacation days has he earned?

____

5. Fifteen people are ordering breakfast at David's restaurant. Each person orders 2 eggs. How many egg cartons that each holds 1 dozen eggs should David open? How many eggs will he have left over?

____

6. There will be 17 people at Lucia's party. She estimates that each person will eat 2 frankfurters. If frankfurters come 8 to a package, how many packages should Lucia buy?

____

7. Deven has 81 postcards. Each page of his album can hold 10 postcards. How many pages of his album can he fill?

____

8. Ellen has 37 ceramic frogs. At most, she can put 7 frogs on each shelf of a display case. How many shelves does she need to display all her frogs?

____

Use with page 25.

Problem-Solving Strategy: Missing Information

Name ______________________

Date ______________________

Cadao buys some school supplies at Perkins Stationery Store during their sale. He buys a notebook for $3.98, an electric pencil sharpener for $4.39, and a pin-point pen. He gives the clerk a 20-dollar bill. How much change should he receive?

SPECIAL SALE

Item	Price
Mechanical pencils	$1.99 each
Pin-point pens . . .	$ .59 each
Plastic rulers	$1.19 each
Paper clips	$1.89 per box
Pencil erasers . . .	$ .39 for two
Ruled tablets	$1.29 each

There is missing information.
Find it on the poster.

$$\begin{array}{r} \$3.98 \\ 4.39 \\ +\ .59 \\ \hline \$8.96 \end{array} \qquad \begin{array}{r} \$20.00 \\ -\ 8.96 \\ \hline \$11.04 \end{array}$$

Cadao should receive $11.04.

Solve. Find missing information in the poster above.
Do your work on a separate sheet of paper.

1. Bian buys a mechanical pencil and 2 ruled tablets. She also buys magic tape for $1.69. How much does she spend in all?

   ____

2. Charlene's father gives her $5.00 to buy 2 boxes of paper clips. How much change should she bring home?

   ____

3. Index cards cost $3.90 per 100 at Perkins Stationery Store. What would be the total cost of 200 index cards and a pin-point pen?

   ____

4. Mechanical pencils are on sale at Kellers Stationery Store at 3 for $6.00. Where would you pay less for 1 mechanical pencil, at Kellers or Perkins?

   ____

5. Which costs more at Perkins Stationery Store, 10 pencil erasers or 4 pin-point pens? How much more?

   ____

6. Samantha bought 3 plastic rulers for $4.47. How much would she have saved if she had bought them at Perkins store?

   ____

7. Yoko earned $5.00. She wants to use it to buy ruled tablets. At most, how many ruled tablets can she buy at Perkins store?

   ____

8. Theo buys 3 boxes of paper clips at Perkins store. He also buys a box of computer diskettes for $8.95. How much does he spend on paper clips?

   ____

Use with page 26.

Problem-Solving Strategy: More Than One Solution

Name ____________________

Date ____________________

Russ multiplied a single-digit number times itself.
He said that the product is greater than 15 and less than 30.
What number did he use?

To find the number, list the single-digit numbers.
Below each, list the product of the number times itself.

Number	0	1	2	3	4	5	6	7	8	9
Product	0	1	4	9	16	25	36	49	64	81

Look for a product that is greater than 15 and less than 30.
There are two. So there are two possible answers.
Russ used either 4 or 5.

Solve. Do your work on a separate sheet of paper.

1. The sum of a number added to itself is less than 15 and greater than 10. What is the number?

____

2. The difference between a number and 2 times the number is less than 5. The number is greater than 0. What is the number?

____

3. The sum of a 2-digit number and a 1-digit number is less than 100 and greater than 98. What are the numbers?

____

4. The product of a 2-digit number and a 1-digit number other than 1 is greater than 20 and less than 25. What are the numbers?

____

5. The quotient of a 2-digit even number divided by a 1-digit number is a whole number greater than 4 and less than 6. What are the numbers?

____

6. This 2-digit number is an even number. When it is divided by 5, the quotient is a 1-digit even number. What is the 2-digit number?

____

7. Tom said, "I am thinking of a 2-digit number. The tens digit is 3 times the ones digit." What number might Tom be thinking of?

____

8. Sara said, "I am thinking of a 2-digit number. The tens digit is $\frac{1}{2}$ the ones digit." What number might Sara be thinking about?

____

 Use with page 27.

Thousandths

Name ____________________

Date ____________________

$\frac{8}{1000} = 0.008$

Standard Form: 0.008

Word Name: eight thousandths

$\frac{162}{1000} = 0.162$

Standard Form: 0.162

Word Name: One hundred sixty-two thousandths

Write as a decimal.

1. $\frac{6}{1000}$ ________
2. $\frac{78}{1000}$ ________
3. $\frac{407}{1000}$ ________
4. $\frac{50}{1000}$ ________

Write the value of the underlined digit.

5. 0.$\underline{6}$13 ________
6. 0.2$\underline{5}$7 ________
7. 0.0$\underline{9}$1 ________
8. 0.20$\underline{6}$ ________

Write the decimal in standard form.

9. thirty-two thousandths ________
10. one thousandth ________
11. nine thousandths ________
12. two hundred nine thousandths ________
13. six hundred twelve thousandths ________
14. fifty thousandths ________

Write the word name for each decimal.

15. 0.941 ____________________
16. 0.007 ____________________
17. 0.086 ____________________
18. 0.301 ____________________
19. 0.040 ____________________
20. 0.800 ____________________

PROBLEM SOLVING

21. A metal rod is six hundred twenty-five thousandths of a meter long. Write this length as a decimal in standard form. ________

Use with Lesson 1-4, text pages 36–37.

Decimals Greater Than One

Name ______________________

Date ______________________

Ones	Tenths	Hundredths	Thousandths
5.	3	7	4

Standard Form: 5.374

Word Name: five *and* three hundred seventy-four thousandths

Write the place of the underlined digit. Then write its value.

1. $4.\underline{8}32$ ______________________
2. $61.67\underline{2}$ ______________________
3. $106.2\underline{4}5$ ______________________
4. $1\underline{5}.133$ ______________________
5. $\underline{2}28.7$ ______________________
6. $\underline{9}4.01$ ______________________

Write each number in standard form.

7. seventy-nine and four hundred thirty-one thousandths ______________
8. two hundred three and six tenths ______________
9. five and eighty-eight hundredths ______________
10. nine hundred ninety-nine and four thousandths ______________
11. three and fifty-two thousandths ______________

Write the word name for each number.

12. 16.72 ______________________________________
13. 4.285 ______________________________________
14. 210.009 ______________________________________
15. 58.007 ______________________________________
16. 116.8 ______________________________________
17. 34.34 ______________________________________
18. 8.031 ______________________________________

PROBLEM SOLVING

19. Natalie walked a mile in 19.086 minutes. Write her time in words. ______________________

20. Ben's time for the one-mile walk was seventeen and five hundred two thousandths minutes. Write Ben's time in standard form. ______________________

 Use with Lesson 1-5, text pages 38–39.

Compare and Order Numbers

Name ______________________________

Date ______________________________

Order from least to greatest: 3581, 3851, 3158, 3285

Align by place value.

3**5**81
3**8**51
3**1**58
3**2**85

Compare the digits in each place, starting at the left.

3 = 3
1 < 2, 2 < 5, 5 < 8

The order from least to greatest: 3**1**58, 3**2**85, 3**5**81, 3**8**51

To compare and order decimals, use the same rules for comparing and ordering whole numbers.

Compare. Write <, =, or >.

1. 47,693 ______ 47,764
2. 78,240 ______ 100,299
3. 675,167 ______ 675,157
4. 0.45 ______ 4.5
5. 0.02 ______ 0.020
6. 5.842 ______ 0.584
7. 8,621,905 ______ 8,621,935
8. 937,435,010 ______ 937,350,119
9. 6,054,835,199 ______ 654,835,199
10. 24,009,075 ______ 240,090,750

Write in order from least to greatest.

11. 92,248; 93,248; 93,148; 94,000 ______________________

12. 612,038; 621,038; 622,037; 612,037 ______________________

13. 7,835,620; 7,835,590; 7,825,780; 783,590 ______________________

14. 38; 0.38; 3.8 ______________________

15. 0.01; 0.1; 0.001 ______________________

16. 2.213; 2.243; 2.231 ______________________

17. 0.67; 0.668; 0.68 ______________________

Write in order from greatest to least.

18. 3265; 327; 3270; 3720 ______________________

19. 11,450; 111,450; 111,540; 1145 ______________________

20. 6,974,000; 6,447,000; 6,947,000; 699,999 ______________________

21. 3.58; 0.358; 35.8 ______________________

22. 9.70; 9.07; 90.7 ______________________

23. 1.063; 1.633; 1.033 ______________________

24. 21.240; 21.426; 21.264 ______________________

Rounding Numbers

Name ______________________

Date ______________________

Round to the place of the underlined digit.

5<u>4</u>,906 ⟶ 55,000 | 3.<u>5</u>16 ⟶ 3.5
<u>6</u>52,163 ⟶ 700,000 | 7<u>1</u>.571 ⟶ 72
8,4<u>0</u>3,177 ⟶ 8,400,000 | $<u>8</u>64.12 ⟶ $900.00

Round each number to the place of the underlined digit.

1. 2<u>9</u>1,564 ____________
2. 67<u>8</u>,519 ____________
3. 8<u>4</u>5,009 ____________
4. 3<u>7</u>2,509,387 ____________
5. 95<u>4</u>,618,085 ____________

Round each to the greatest place.

6. 37,121 ____________
7. 507,892 ____________
8. 495,382 ____________
9. 254,068 ____________
10. 24,491,630 ____________
11. 85,037,124 ____________

Round each to the nearest whole number, tenth, and hundredth.

12. 5.419 ____________
13. 8.566 ____________
14. 67.935 ____________
15. 22.193 ____________
16. 84.177 ____________
17. 39.528 ____________
18. 128.726 ____________
19. 434.582 ____________
20. 275.509 ____________

Round each to the nearest ten cents, dollar, ten dollars, and hundred dollars.

21. $815.76 ____________
22. $352.68 ____________
23. $549.53 ____________
24. $922.58 ____________
25. $406.39 ____________
26. $634.82 ____________

Use with Lesson 1-7, text pages 42–43.

Addition Properties/ Subtraction Rules

Name ______________________

Date ______________________

Commutative Property	Identity Property	Associative Property
7 + 6 = 13 6 + 7 = 13	5 + 0 = 5 0 + 5 = 5	(3 + 4) + 8 = 15 3 + (4 + 8) = 15
5 + 4 = 9 9 − 4 = 5 — Inverse Operations	4 − 4 = 0 4 − 0 = 4	

Write the missing number. Name the property of addition that is used.

1. 9 + 3 = ___ + 9 ______________ **2.** 7 = 7 + ___ ______________

3. 4 = ___ + 4 ______________ **4.** 8 + ___ = 5 + 8 ______________

5. 7 + (6 + 3) = (7 + 6) + ___ ______________

6. (5 + ___) + 3 = 5 + (4 + ___) ______________

Add. Use properties of addition to find shortcuts.

7. 1 + 3 + 7 + 5 = ___

8. 4 + 1 + 6 + 3 = ___

9. 9 + 2 + 1 + 6 = ___

10. 0 + 8 + 2 + 9 = ___

11. 6 + 7 + 4 + 0 = ___

12. 3 + 5 + 2 + 5 = ___

13. 4 + 7 + 6 + 3 = _____ **14.** 1 + 5 + 6 + 7 = _____ **15.** 0 + 5 + 7 + 2 = _____

16. 1 + 8 + 1 + 9 = _____ **17.** 2 + 3 + 4 + 5 = _____ **18.** 1 + 0 + 9 + 5 = _____

Complete.

19. 2 + 6 = 8
8 − 6 = ___

20. 9 − 5 = 4
4 + ___ = 9

21. 9 + 7 = 16
16 − ___ = 9

22. 1 − 1 = 0
0 + 1 = ___

23. 16 − 9 = 7
___ + 9 = 16

24. 7 + 4 = 11
___ − 4 = 7

Write the missing addend.

25. 8 + ____ = 17 **26.** ____ + 9 = 11 **27.** 7 + ____ = 13

28. 6 + ____ = 14 **29.** ____ + 0 = 4 **30.** ____ + 9 = 10

31. ____ + 8 = 8 **32.** 5 + ____ = 13 **33.** 2 + ____ = 9

Estimating Sums and Differences

Name ______________________

Date ______________________

Front-End Estimation			**Rounding**		
	5263	about 1000		5263 →	5300
	2886			2886 →	2900
	694	about 1000		694 →	700
	+ 7243			+ 7243 →	+ 7200
rough estimate: **14**,000					16,100
adjusted estimate: 14,000 + 1000 + 1000 = 16,000					

Estimate the sum. Use front-end estimation. Then use rounding.

1. 115 + 173

2. 48 + 714

3. 533 + 561

4. 216 + 468

5. 746 + 752 + 398

6. 903 + 746 + 88

7. 634 + 71 + 15

8. 431 + 401 + 275

9. 9925 + 8395 + 4093

10. 8052 + 9894 + 2584

11. $17.44 + 69.46 + 1.69

12. $33.90 + 69.25 + 11.03

13. 6327 + 4896 + 2929 = ______________

14. $81.17 + $19.87 + $92.66 = ______________

Estimate the difference. Use front-end estimation. Then use rounding.

15. 857 − 723

16. $9.93 − 4.41

17. 415 − 73

18. $6.51 − 2.98

19. 5992 − 4069

20. 9450 − 6846

21. $59.14 − 13.19

22. 6150 − 3891

23. 8792 − 7128 = __________

24. $63.17 − $14.92 = __________

PROBLEM SOLVING

25. On a business trip a salesperson drove 372 miles the first day, 338 miles the second day, and 178 miles the third day. Estimate her total mileage. ______________

Use with Lesson 1-9, text pages 46–47.

Addition: Three or More Addends

Name ______________________

Date ______________________

Add: 3476 + 409 + 1824 = ?

Estimate.

3476	→	3500
409	→	400
+ 1824	→	+ 1800
		about 5700

Add.

$$\begin{array}{r} {\scriptstyle 1\,1\,1} \\ 3476 \\ 409 \\ +\ 1824 \\ \hline 5709 \end{array}$$

Estimate. Then add.

1.	2.	3.	4.	5.
96 87 + 51	94 36 + 62	76 5 + 59	533 106 + 999	695 980 + 243

6.	7.	8.	9.	10.
100 103 + 798	676 15 + 575	284 45 + 295	382 300 + 948	7852 3789 + 2896

11.	12.	13.	14.	15.
7852 6317 + 6276	4284 979 + 5610	6944 9137 + 8348	102 1516 + 3774	1881 9795 + 8721

16.	17.	18.	19.	20.
\$1.40 6.07 + 8.63	\$81.44 70.90 + 52.64	\$ 4.53 41.40 + 9.39	\$24.38 97.90 + 6.09	\$39.02 36.49 + 48.98

21.	22.	23.	24.	25.
22 543 916 + 430	677 502 513 + 462	2438 9790 609 + 3902	3534 4048 2697 + 555	\$ 1.59 7.24 10.03 + 45.45

Align and add.

26. 468 + 298 + 622 = __________

27. 3019 + 7747 + 140 + 7607 = __________

28. 67 + 263 + 279 = __________

29. 8626 + 7904 + 3735 + 9619 = __________

30. 695 + 980 + 243 = __________

31. 7055 + 5334 + 5795 + 2895 = __________

32. 829 + 24 + 589 = __________

33. 8144 + 7090 + 453 + 4140 = __________

34. 86 + 910 + 226 = __________

35. 8714 + 562 + 9495 + 3640 = __________

Subtraction with Zeros

Name ______________________

Date ______________________

To subtract when the minuend has zeros, regroup as many times as necessary *before* you start to subtract.

```
    9  9
 7 10 10 10
  8  0  0  0
  6  4  9  3
 -----------
  1  5  0  7
```

Subtract and check.

1. 60 − 25	**2.** 80 − 17	**3.** 50 − 32	**4.** 40 − 21	**5.** 90 − 67
6. 200 − 84	**7.** 500 − 314	**8.** 300 − 158	**9.** \$7.00 − 2.98	**10.** \$6.00 − 4.34
11. 7000 − 6193	**12.** 4000 − 2864	**13.** 9000 − 5877	**14.** 5000 − 1891	**15.** 8000 − 4375
16. 1006 − 729	**17.** 3004 − 1949	**18.** 2001 − 1863	**19.** 6008 − 3855	**20.** 8005 − 4466
21. 8040 − 6192	**22.** \$60.00 − 26.39	**23.** \$40.50 − 38.76	**24.** \$50.02 − 9.87	**25.** \$90.07 − 7.58

Align and subtract.

26. 8000 − 7638 = ____________ **27.** \$60.03 − \$27.95 = ____________

28. 2070 − 999 = ____________ **29.** \$80.00 − \$16.27 = ____________

30. 7004 − 1928 = ____________ **31.** \$50.20 − \$7.68 = ____________

PROBLEM SOLVING

32. Shaya saved \$50.00. She bought a jacket that cost \$27.39. Did she have enough money left to buy a skirt for \$23.99? ______________________

 Use with Lesson 1-11, text pages 50–51.

Larger Sums and Differences

Name ______________________

Date ______________________

Add: 696,285 + 401,604 + 16,297 = ?

Estimate.

```
696,285 ——→   700,000
401,604 ——→   400,000
+ 16,297 ——→ + 20,000
       about 1,120,000
```

Add.

```
   111 12
  696,285
  401,609
+  16,297
1,114,191
```

Subtract: $1860.13 − $1085.01 = ?

Estimate.

```
$1860.13 ——→   $2000.00
− 1085.01 ——→ − 1000.00
         about $1000.00
```

Subtract.

```
    15
   7 8 10
 $1860.13
− 1085.01
 $ 775.12
```

Estimate. Then add or subtract.

1. 95,127 + 54,202

2. 49,437 + 97,296

3. 21,779 + 37,899

4. 64,595 + 34,783

5. 39,584 + 28,150

6. 50,337 + 13,873

7. 51,725 + 96,536

8. 10,443 + 18,540

9. $862.61 + 790.48

10. $373.53 + 961.95

11. $871.44 + 56.23

12. $545.01 + 419.70

13. 42,845 − 4,197

14. 56,104 − 38,963

15. 91,371 − 76,278

16. 57,794 − 34,063

17. $155.66 − 73.85

18. $257.26 − 161.75

19. $542.07 − 84.72

20. $911.19 − 723.67

21. 70,544 + 53,342 + 57,951

22. 28,956 + 30,194 + 77,474

23. 1,401 + 76,072 + 81,449

24. 70,903 + 4,535 + 41,403

25. 972,729 + 804,267 + 674,826

26. 905,129 + 875,779 + 416,648

27. 123,082 + 954,157 + 797,297

28. $1692.64 + 7127.30 + 2324.96

Align. Then add or subtract.

29. 518,015 + 757,029 = ______________

30. 278,279 − 44,647 = ______________

31. 784,488 + 50,262 = ______________

32. 162,821 − 105,277 = ______________

Roman Numerals

Name ____________________

Date ____________________

I	II	III	IV	V	VI	VII	VIII	IX	X
1	2	3	4	5	6	7	8	9	10
V	X	XV	XX	XXV	XXX	XXXV	XL	XLV	L
5	10	15	20	25	30	35	40	45	50
X	XX	XXX	XL	L	LX	LXX	LXXX	XC	C
10	20	30	40	50	60	70	80	90	100
C	CC	CCC	CD	D	DC	DCC	DCCC	CM	M
100	200	300	400	500	600	700	800	900	1000

Complete.

1. MCLXXV = 1000 + ____ + 50 + ____ + ____ + 5 = ____

2. CMXLIX = (1000 − ____) + (____ − 10) + (____ − ____) = ____

3. CDXXXIV = (____ − 100) + (____ + ____ + ____) + (____ − ____) = ____

Write the Roman numeral in standard form.

4. MCDXL ____________ **5.** CDIII ____________ **6.** XCIV ____________

7. CMVI ____________ **8.** MCMXLVII ____________ **9.** MMXX ____________

10. CMXC ____________ **11.** CMXLV ____________ **12.** DCCXX ____________

13. CCLXIV ____________ **14.** XLIX ____________ **15.** LXVIII ____________

Write each as a Roman numeral.

16. 54 ____________ **17.** 1985 ____________ **18.** 3009 ____________

19. 480 ____________ **20.** 1507 ____________ **21.** 2645 ____________

22. 1535 ____________ **23.** 729 ____________ **24.** 1006 ____________

25. 665 ____________ **26.** 333 ____________ **27.** 147 ____________

Write the date each state was admitted into the Union as a Roman numeral.

28. Delaware – 1787 ____________ **29.** Idaho – 1890 ____________ **30.** Alaska – 1959 ____________

PROBLEM SOLVING

31. The year the United States Constitution was ratified is written in Roman numerals as MDCCLXXXIX. What is the standard number for this year? ____________

Problem-Solving Strategy: Guess and Test

Name ______________________________

Date ______________________________

Kerry has $15 less than Kevin. Together they have $45. How much money does each person have?

Guess an amount of money for Kevin.
Subtract $15 to find the amount for Kerry.
Check to see whether the sum is $45.

	Guess 1	Guess 2	Guess 3
Kevin	$40	$20	$30
Kerry	$25	$5	$15
Sum	$65	$25	$45

Kevin has $30. Kerry has $15.

Solve. Do your work on a separate sheet of paper.

1. Michelle and Michael collect baseball cards. Michael has 5 more cards than Michelle. Together they have 27 cards. How many cards does each have?

   ____

2. There are 12 red and blue pencils on a desk. There are twice as many red pencils as blue pencils. How many red pencils are on the desk?

   ____

3. Elena is 6 years younger than her brother Geraldo. The sum of their ages is 30. How old is Elena?

   ____

4. Doug paid $22 for two books. One book cost $5 more than the other. How much did each book cost?

   ____

5. Jeremy and Hanna collect coins. Jeremy has 7 fewer coins than Hanna. If together they have 83 coins, how many does Hanna have?

   ____

6. Linda received 75¢ in change, all nickels and dimes. There are 11 coins. How many nickels and how many dimes did she receive?

   ____

7. Jamie wants to get a fox, a goose, and a bag of grain safely across a river in a boat. He can take only 1 of the 3 in his boat at one time. If left alone together, the goose would eat the grain and the fox would eat the goose. How many trips will Jamie need to make?

   ____

8. An animal trainer wants to move 2 lions and a chimpanzee to another training area. He can transport only one animal at a time. He cannot leave the chimpanzee alone with either lion. How many trips will he have to make?

   ____

Meaning and Properties of Multiplication

Name ______________________

Date ______________________

Commutative Property	$5 \times 7 = 7 \times 5 = 35$
Associative Property	$(2 \times 7) \times 3 = 2 \times (7 \times 3) = 42$
Identity Property	$9 \times 1 = 1 \times 9 = 9$
Zero Property	$3 \times 0 = 0 \times 3 = 0$
Distributive Property	$5 \times (4 + 3) = (5 \times 4) + (5 \times 3) = 35$

Write the multiplication fact.

1. $2 + 2 + 2$ ______________ **2.** $5 + 5 + 5$ ______________

3. $1 + 1 + 1 + 1$ ______________ **4.** $9 + 9 + 9 + 9 + 9$ ______________

5. 7 ______________ **6.** $8 + 8 + 8 + 8$ ______________

Write the name of each property of multiplication.

7. $9 \times 7 = 7 \times 9$ ______________ **8.** $1 \times 0 = 0$ ______________

9. $2 \times (6 \times 3) = (2 \times 6) \times 3$ ______________

10. $2 \times (5 + 2) = (2 \times 5) + (2 \times 2)$ ______________

11. $3 \times 1 = 3$ ______________ **12.** $0 \times 6 = 6$ ______________

Write the missing number.

13. ___ $\times 3 = 3 \times 5$ **14.** $2 \times (7 + 8) = (2 \times$ ___$) + (2 \times 8)$

15. $8 \times$ ___ $= 0$ **16.** $8 \times 5 =$ ___ $\times 8$

17. $7 \times$ ___ $= 7$ **18.** $6 \times (7 \times 8) = (6 \times$ ___$) \times 8$

19. ___ $\times 4 = 4$ **20.** $3 \times (1 + 5) = ($___ $\times 1) + (3 \times 5)$

21. $9 \times$ ___ $= 36$ **22.** $2 \times$ ___ $= 2$ **23.** $5 \times$ ___ $= 45$ **24.** $3 \times$ ___ $= 24$ **25.** ___ $\times 8 = 16$ **26.** $4 \times$ ___ $= 28$

PROBLEM SOLVING

27. The product is 72. One factor is 9. What is the other factor? ______________

28. There are 5 black keys in each octave on a piano. How many black keys are there on a 6-octave piano? ______________

29. In a classroom there are 7 rows of desks with 4 desks in each row. How many desks are there in the classroom? ______________

Mental Math: Special Factors

Name ______________________

Date ______________________

Look for patterns when you multiply with 10 and with multiples of 10.

10 × 1 = 10	50 × 4 = 200	9 × 20 = 180 (one zero)	80 × 5 = 400 (one zero)
10 × 2 = 20	50 × 5 = 250		
10 × 3 = 30	50 × 6 = 300		
10 × 4 = 40	50 × 7 = 350		

Write the products.

1. 10 × 5 = ______
   10 × 6 = ______
   10 × 7 = ______

2. 2 × 30 = ______
   2 × 40 = ______
   2 × 50 = ______

3. 30 × 6 = ______
   30 × 7 = ______
   30 × 8 = ______

4. 3 × 60 = ______
   4 × 60 = ______
   5 × 60 = ______

5. 50 × 7 = ______
   50 × 8 = ______
   50 × 9 = ______

6. 7 × 70 = ______
   7 × 80 = ______
   7 × 90 = ______

Multiply.

7. 60 × 2
8. 80 × 6
9. 70 × 5
10. 3 × 50
11. 9 × 30
12. 8 × 40

13. 10 × 9
14. 90 × 4
15. 20 × 7
16. 6 × 60
17. 7 × 80
18. 9 × 70

19. 3 × 50 = ______
20. 20 × 6 = ______
21. 40 × 5 = ______
22. 70 × 6 = ______
23. 9 × 40 = ______
24. 8 × 60 = ______

PROBLEM SOLVING

25. Raynell opened 2 crates of oranges. Each crate contained 10 bags of oranges. There were 6 oranges in each bag. How many oranges were in the crates? ______________

26. Lisa placed 10 cards in each of 5 rows. How many cards were there? ______________

Use with Lesson 2-3, text pages 70–71.

Patterns in Multiplication

Name ______________________

Date ______________________

Look for patterns when you multiply with 100, 1000, or their multiplies.

5 × 6 = 30
50 × 6 = 300
500 × 6 = 3000
5000 × 6 = 30,000

50 × 60 = 3000
500 × 60 = 30,000
5000 × 60 = 300,000

600 × 7 = 4200 (2 zeros)

70 × 3000 = 210,000 (4 zeros)

Multiply.

1. 60 × 20 = ______
600 × 20 = ______
6000 × 20 = ______

2. 50 × 5 = ______
500 × 5 = ______
5000 × 5 = ______

3. 80 × 40 = ______
800 × 40 = ______
8000 × 40 = ______

4. 70 × 6 = ______
700 × 6 = ______
7000 × 6 = ______

5. 50 × 70 = ______
500 × 70 = ______
5000 × 70 = ______

6. 30 × 90 = ______
300 × 90 = ______
3000 × 90 = ______

7. 10 × 10 = ______
100 × 10 = ______
1000 × 10 = ______

8. 20 × 60 = ______
200 × 60 = ______
2000 × 60 = ______

9. 80 × 80 = ______
800 × 80 = ______
8000 × 80 = ______

10. 6 × 300

11. 8 × 200

12. 7 × 5000

13. 5 × 6000

14. 4 × 9000

15. 10 × 800

16. 20 × 500

17. 30 × 4000

18. 90 × 2000

19. 80 × 5000

20. 600 × 6 = ______

21. 300 × 90 = ______

22. 800 × 30 = ______

23. 7 × 5000 = ______

24. 30 × 3000 = ______

25. 50 × 4000 = ______

26. 1000 × 50 = ______

27. 9 × 7000 = ______

28. 7000 × 20 = ______

PROBLEM SOLVING

29. If each of 8000 runners carries a torch for 6 kilometers before passing it on to the next runner, how much distance do they cover? ______

Estimating Products

Name ______________________________

Date ______________________________

To estimate the product of two numbers:
- Round each factor to its greatest place.
- Multiply.

$$\begin{array}{r} 217 \\ \times\ 189 \\ \hline \end{array} \longrightarrow \begin{array}{r} 200 \\ \times\ 200 \\ \hline \text{about } 40{,}000 \end{array} \qquad \begin{array}{r} 787 \\ \times\ 934 \\ \hline \end{array} \longrightarrow \begin{array}{r} 800 \\ \times\ 900 \\ \hline \text{about } 720{,}000 \end{array}$$

Estimate each product.

1. 44×27
2. 87×75
3. 27×67
4. 79×29
5. 23×48
6. 25×34
7. 223×58
8. 363×87
9. 478×19
10. 684×74
11. 613×38
12. 161×80
13. 202×95
14. 793×47
15. 463×11
16. $\$8.75 \times 4$
17. $\$95.17 \times 13$
18. $\$59.07 \times 52$
19. $\$89.49 \times 821$
20. $\$17.24 \times 108$

Estimate the sum. Use clustering.

21. 39 + 42 + 40 + 38 + 43 + 41 = ______________
22. 78 + 81 + 79 = ______________
23. 192 + 197 + 201 + 212 + 225 = ______________
24. 289 + 299 + 304 + 294 + 298 + 315 = ______________
25. 65 + 73 + 70 + 69 + 68 + 74 + 71 = ______________

PROBLEM SOLVING

26. A certain bicycle costs $199.95. About how much would 75 of these bicycles cost? ______________

27. An average chicken weighs 7 pounds. The average cost per pound is $1.09. About how much would 200 chickens cost? ______________

28. Sneakers are on sale for $49.99 per pair. About how much would 50 pairs cost? ______________

Zeros in the Multiplicand

Name ____________________

Date ____________________

Multiply: 6 × 4507 = ?

Multiply ones. Regroup. Multiply tens. Add.

$$\begin{array}{r} \scriptstyle 4 \\ 4507 \\ \times \quad 6 \\ \hline 42 \end{array}$$

6 × 7 ones = 4 tens 2 ones.
6 × 0 tens = 0 tens.
0 tens + 4 tens = 4 tens.

Multiply hundreds. Regroup. Multiply thousands. Add.

$$\begin{array}{r} \scriptstyle 3\ \ 4 \\ 4507 \\ \times \quad 6 \\ \hline 27{,}042 \end{array}$$

6 × 5 hundreds = 3 thousands 0 hundreds.
6 × 4 thousands = 24 thousands.
24 thousands + 3 thousands = 27 thousands.

Estimate. Then multiply.

1. 608 × 4
2. 309 × 9
3. 705 × 7
4. 8043 × 4
5. 4061 × 8
6. 5075 × 5
7. 2099 × 6
8. 9408 × 2
9. 6705 × 3
10. 8904 × 7
11. 7008 × 5
12. 92,605 × 6
13. 44,031 × 8
14. 20,603 × 9
15. 50,987 × 2
16. 40,090 × 3
17. 30,008 × 6
18. 10,800 × 4
19. 35,007 × 7
20. 70,006 × 8

Write the product. You may use the distributive property.

21. 5 × 3087 = ____________
22. 2 × 9708 = ____________
23. 6 × 40,504 = ____________
24. 9 × 10,056 = ____________
25. 4 × 71,002 = ____________
26. 3 × 90,085 = ____________
27. 9 × 73,000 = ____________
28. 7 × 80,310 = ____________

PROBLEM SOLVING

29. Each of the 4 sides of a building was faced with 25,080 bricks. How many bricks were used to face all the sides of the building? ____________

 Use with Lesson 2-6, text pages 76–77.

Multiplying Two Digits

Name ______________________

Date ______________________

To multiply by two digits:			
• Multiply by the ones.	23 × 13	346 × 54	4721 × 76
• Multiply by the tens.	69	1384	28326
• Add the partial products.	23	1730	33047
	299	18,684	358,796

Multiply.

1. 10 × 78	**2.** 45 × 45	**3.** 59 × 83	**4.** 16 × 61	**5.** 20 × 95
6. 454 × 35	**7.** 149 × 52	**8.** 928 × 53	**9.** 89 × 75	**10.** 401 × 46
11. 645 × 34	**12.** 104 × 18	**13.** 77 × 23	**14.** 959 × 84	**15.** 494 × 97
16. 9967 × 33	**17.** 4952 × 41	**18.** 6952 × 17	**19.** 4229 × 64	**20.** 8316 × 23

Write the product.

21. 54 × 8146 = ______________

22. 70 × 4275 = ______________

23. 61 × 2277 = ______________

24. 28 × 4898 = ______________

25. 80 × 6985 = ______________

26. 98 × 2232 = ______________

PROBLEM SOLVING

27. Growing Gardens had 18 daylily plants. Each plant produced 875 flowers. How many flowers did the plants produce? ______________

 Use with Lesson 2-7, text pages 78–79.

Multiplying Three Digits

Name ______________________

Date ______________________

To multiply by three digits:
- Multiply by the ones.
- Multiply by the tens.
- Multiply by the hundreds.
- Add the partial products.

$$\begin{array}{r} 721 \\ \times\ 184 \\ \hline 2884 \\ 5768 \\ 721 \\ \hline 132{,}664 \end{array} \qquad \begin{array}{r} 2753 \\ \times\ 239 \\ \hline 24777 \\ 8259 \\ 5506 \\ \hline 657{,}967 \end{array}$$

Multiply.

1. 218 × 446	**2.** 236 × 878	**3.** 610 × 374	**4.** 397 × 526	**5.** 270 × 581
6. 868 × 259	**7.** 259 × 178	**8.** 585 × 931	**9.** 438 × 759	**10.** 245 × 378
11. 312 × 798	**12.** 151 × 592	**13.** 956 × 243	**14.** 939 × 114	**15.** 984 × 632

Write the product.

16. 9795 × 872 = ____________ **17.** 5815 × 729 = ____________

18. 1450 × 259 = ____________ **19.** 5044 × 183 = ____________

PROBLEM SOLVING

20. A farmer planted 245 rows of cabbages. There were 267 cabbages in each row. How many cabbages did he plant? ____________

21. In one week, Jane picked 623 baskets of cherries. If each basket held 382 cherries, how many cherries did she pick? ____________

 Use with Lesson 2-8, text pages 80–81.

Zeros in the Multiplier

Name ______________________________

Date ______________________________

When multiplying with zeros in the multiplier:

- You may omit the partial products of the zeros.
- Remember to align the other partial products correctly under the multiplier place.

Long Way	Short Way
8241	8241
× 607	× 607
57687	57687
0000	49446
49446	5,002,287
5,002,287	

Multiply.

1. 580 × 605	**2.** 268 × 508	**3.** 406 × 950	**4.** 763 × 580	**5.** 651 × 807
6. 807 × 309	**7.** 598 × 902	**8.** 128 × 650	**9.** 412 × 510	**10.** 733 × 450
11. 3326 × 450	**12.** 5551 × 803	**13.** 5639 × 203	**14.** 2492 × 807	**15.** 6181 × 306
16. 3119 × 240	**17.** 8225 × 209	**18.** 5899 × 602	**19.** 4216 × 504	**20.** 2631 × 390

Write the product.

21. 200 × 785 = ______________

22. 300 × 254 = ______________

23. 600 × 3476 = ______________

24. 900 × 1769 = ______________

25. 500 × 7776 = ______________

26. 700 × 9753 = ______________

PROBLEM SOLVING

27. A stadium has 200 rows of 225 seats. How many seats are in the stadium? ______________

 Use with Lesson 2-9, text pages 82–83.

Multiplication with Money

Name ______________________

Date ______________________

To multiply an amount of money:
- Multiply as usual.
- Write a decimal point in the product two places from the right.
- Write the dollar sign in the product.

```
   $.59          $7.87
 ×   31        ×   264
     59           3148
   177           4722
 $18.29         1574
              $2077.68
```

Multiply.

1. $0.09 × 3	**2.** $1.02 × 4	**3.** $2.22 × 5	**4.** $6.12 × 9	**5.** $4.95 × 2
6. $1.47 × 8	**7.** $0.36 × 9	**8.** $7.92 × 6	**9.** $0.82 × 4	**10.** $1.66 × 7
11. $8.36 × 23	**12.** $4.92 × 79	**13.** $6.30 × 43	**14.** $.83 × 98	**15.** $7.80 × 55
16. $7.56 × 70	**17.** $3.81 × 33	**18.** $0.87 × 107	**19.** $.56 × 60	**20.** $2.45 × 598
21. $1.50 × 792	**22.** $2.21 × 445	**23.** $6.66 × 853	**24.** $4.99 × 141	**25.** $7.42 × 276

Write the product.

26. 810 × $2.72 = ______________ **27.** 210 × $3.02 = ______________

PROBLEM SOLVING

28. How much would 75 gallons of gasoline cost if the price is $1.29 per gallon? ______________

Problem-Solving Strategy: Hidden Information

Name ______________________

Date ______________________

Harrison reads about 3 novels every month. At this rate, about how many novels does he read in 2 years?

You need to know the number of months in a year.
There are 12 months in a year.
So in 2 years there are 2 × 12 or 24 months.

Harrison reads about 24 × 3 or 72 novels in 2 years.

Solve. Do your work on a separate sheet of paper.

1. Kari exercises by walking at a fast pace. She walks about 3 miles every weekday. How many miles does Kari walk in 6 weeks?

____

2. The Ramirez family leaves on a vacation trip July 24. They return 14 days later. On what date does the family return from vacation?

____

3. Keisha can read 5 pages of her book in 15 minutes. If she reads for 2 hours, how many pages will she read?

____

4. Each side of a pentagon is 37 cm long. What is the distance around the pentagon?

____

5. Suni watches 3 television shows during the weekend. Each show lasts a half hour. How much time does she spend watching television during the weekend?

____

6. The chef at the Omelet Magic has 5 dozen eggs. He uses 4 eggs to make an omelet. How many eggs does he have left?

____

7. Danny uses 2 measures of fish food in each fish tank in the pet store every day. There are 37 fish tanks. How many measures of fish food does Danny use in one week?

____

8. Pablo spends $9 every week on bus fare. How much money will he spend on bus fare in one year?

____

9. Mary has $2\frac{1}{2}$ ft of ribbon. She needs 36 in. of ribbon for a craft project. How much more ribbon does she need?

____

10. Ben needs 3 cups of milk for a recipe. He has 1 pint of milk in the refrigerator. How much more milk does he need?

____

Understanding Division

Name ______________________

Date ______________________

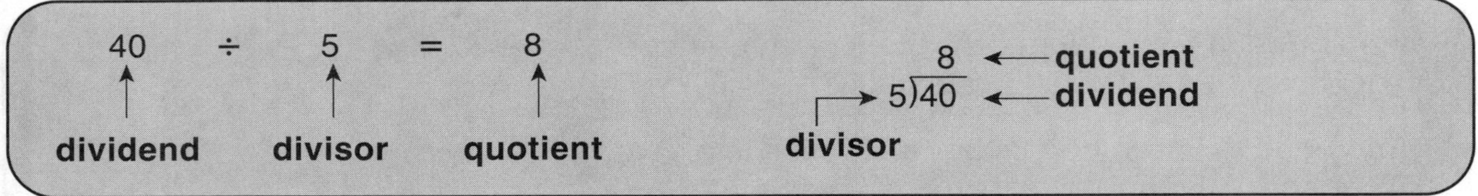

Complete.

1. 4 × 5 = 20
20 ÷ 5 = ___
20 ÷ 4 = ___

2. 7 × 6 = 42
42 ÷ ___ = 6
___ ÷ 6 = ___

3. 9 × 6 = 54
54 ÷ ___ = ___
54 ÷ ___ = ___

4. ___ × ___ = 72
72 ÷ 8 = ___
___ ÷ ___ = ___

Write all the related facts using the given numbers.

5. 4, 6, 24
___ × ___ = ___
___ × ___ = ___
___ ÷ ___ = ___
___ ÷ ___ = ___

6. 7, 7, 49
___ × ___ = ___
___ ÷ ___ = ___

7. 5, 3, 15
___ × ___ = ___
___ × ___ = ___
___ ÷ ___ = ___
___ ÷ ___ = ___

8. 4, 9, 36
___ × ___ = ___
___ × ___ = ___
___ ÷ ___ = ___
___ ÷ ___ = ___

9. 7, 9, 63
___ × ___ = ___
___ × ___ = ___
___ ÷ ___ = ___
___ ÷ ___ = ___

10. 6, 5, 30
___ × ___ = ___
___ × ___ = ___
___ ÷ ___ = ___
___ ÷ ___ = ___

11. 3, 8, 24
___ × ___ = ___
___ × ___ = ___
___ ÷ ___ = ___
___ ÷ ___ = ___

12. 2, 6, 12
___ × ___ = ___
___ × ___ = ___
___ ÷ ___ = ___
___ ÷ ___ = ___

Complete.

13. $7\overline{)5}$ **14.** $\overset{9}{\overline{)81}}$ **15.** $4\overline{)28}$ **16.** $8\overline{)56}$ **17.** $3\overline{)27}$

18. $\overset{8}{\overline{)32}}$ **19.** $\overset{5}{\overline{)20}}$ **20.** $4\overline{)16}$ **21.** $5\overline{)5}$ **22.** $\overset{3}{1\overline{)\quad}}$

23. $7\overline{)7}$ **24.** $\overset{0}{9\overline{)\quad}}$ **25.** $6\overline{)36}$ **26.** $8\overline{)64}$ **27.** $6\overline{)48}$

28. $3\overline{)21}$ **29.** $6\overline{)36}$ **30.** $\overset{7}{\overline{)56}}$ **31.** $9\overline{)72}$ **32.** $5\overline{)40}$

33. $\overset{7}{\overline{)42}}$ **34.** $7\overline{)49}$ **35.** $6\overline{)54}$ **36.** $1\overline{)6}$ **37.** $\overset{6}{\overline{)24}}$

Division Patterns

Name ______________________________

Date ______________________________

Use division facts and patterns with zero to divide with multiples of 10, 100, or 1000.

6	÷	**6**	=	**1**	**40**	÷	**8**	=	**5**
60	÷	**6**	=	**1**0	**40**0	÷	**80**	=	**5**
600	÷	**6**	=	**1**00	**40**00	÷	**80**	=	**5**0
6000	÷	**6**	=	**1**000	**40**,000	÷	**80**	=	**5**00
60,000	÷	**6**	=	**1**0,000	**40**0,000	÷	**80**	=	**5**000

Find the quotient. Look for a pattern.

1. 36 ÷ 9 = _____
360 ÷ 9 = _____
3600 ÷ 9 = _____
36,000 ÷ 9 = _____
360,000 ÷ 9 = _____

2. 28 ÷ 7 = _____
280 ÷ 70 = _____
2800 ÷ 700 = _____
28,000 ÷ 7000 = _____
280,000 ÷ 70,000 = _____

3. 54 ÷ 9 = _____
540 ÷ 90 = _____
5400 ÷ 90 = _____
54,000 ÷ 90 = _____
540,000 ÷ 90 = _____

Divide.

4. 6)360 **5.** 8)720 **6.** 7)6300 **7.** 2)1800

8. 30)270 **9.** 40)280 **10.** 50)2500 **11.** 60)4200

12. 70)49,000 **13.** 80)16,000 **14.** 90)270,000 **15.** 20)400,000

16. 3000 ÷ 50 = ________ **17.** 54,000 ÷ 60 = ________ **18.** 28,000 ÷ 70 = ________

Compare. Write <, =, or >.

19. 4200 ÷ 6 ____ 3500 ÷ 7

20. 7200 ÷ 80 ____ 8100 ÷ 90

21. 50,000 ÷ 10 ____ 60,000 ÷ 6

22. 2100 ÷ 70 ____ 2000 ÷ 50

23. 5400 ÷ 9 ____ 3600 ÷ 6

24. 4500 ÷ 5 ____ 5600 ÷ 8

25. 16,000 ÷ 8 ____ 8000 ÷ 4

26. 1600 ÷ 2 ____ 1200 ÷ 4

27. 70,000 ÷ 10 ____ 70,000 ÷ 7

28. 2700 ÷ 3 ____ 3200 ÷ 4

29. 18,000 ÷ 3 ____ 20,000 ÷ 4

30. 2700 ÷ 9 ____ 3000 ÷ 10

31. 40,000 ÷ 4 ____ 50,000 ÷ 5

32. 64,000 ÷ 80 ____ 6400 ÷ 8

Use with Lesson 3-2, text pages 98–99.

Three-Digit Quotients

Name ____________________

Date ____________________

Division Steps
- Decide where to begin the quotient.
- Estimate.
- Divide.
- Multiply.
- Subtract and compare.
- Bring down.
- Repeat the steps as necessary.
- Check.

Divide: 497 ÷ 3 = ?

```
   165  R2
3)497
 - 3↓
   19
 - 18↓
    17
  - 15
     2
```

Check.

```
  165
×   3
  495
+   2
  497
```

Divide and check.

1. $7\overline{)782}$
2. $5\overline{)850}$
3. $6\overline{)672}$
4. $8\overline{)981}$
5. $4\overline{)721}$
6. $2\overline{)695}$
7. $3\overline{)1814}$
8. $5\overline{)933}$
9. $7\overline{)6856}$
10. $4\overline{)3936}$
11. $6\overline{)5849}$
12. $3\overline{)2945}$

PROBLEM SOLVING

13. There are 975 cans at a factory. How many 6-packs can be made? How many cans will be left over?

____

14. Three friends equally share a bag of 813 pennies. How many pennies does each friend get?

____

 Use with Lesson 3-3, text pages 100–101.

Larger Quotients

Name ____________________

Date ____________________

Divide: 9385 ÷ 4 = ?

Repeat the division steps until the division is completed.

$$\begin{array}{r} 2346 \text{ R1} \\ 4\overline{)9385} \\ -8 \\ \hline 13 \\ -12 \\ \hline 18 \\ -16 \\ \hline 25 \\ -24 \\ \hline 1 \end{array}$$

Check.

$$\begin{array}{r} 2346 \\ \times \quad 4 \\ \hline 9384 \\ + \quad 1 \\ \hline 9385 \end{array}$$

Divide and check.

1. $3\overline{)7826}$
2. $5\overline{)8579}$
3. $6\overline{)9672}$
4. $8\overline{)9874}$
5. $6\overline{)43{,}487}$
6. $2\overline{)31{,}543}$
7. $4\overline{)28{,}529}$
8. $9\overline{)61{,}134}$
9. $5\overline{)141{,}585}$
10. $7\overline{)232{,}486}$
11. $8\overline{)522{,}872}$
12. $9\overline{)515{,}056}$

PROBLEM SOLVING

13. Plants Plus has 21,073 tulip bulbs. If 5 bulbs are planted in each pot, how many pots are needed? How many bulbs are left over?

    ____

14. An automobile factory made 8500 cars. The same number of cars were sent to 4 cities. How many cars were sent to each city?

    ____

Zeros in the Quotient

Name ______________________

Date ______________________

5015 ÷ 5 = ?

Not enough hundreds or tens Write 0s in the quotient.

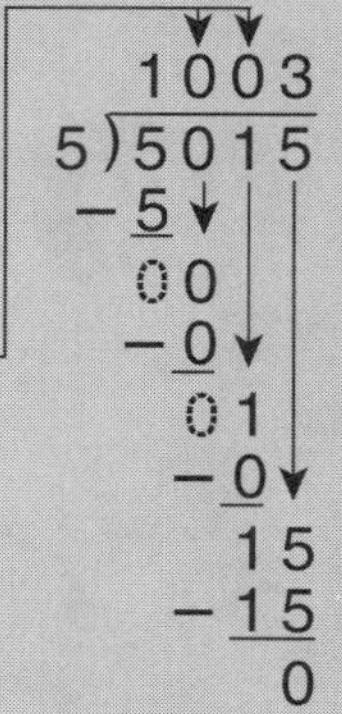

Check.

$$\begin{array}{r} 1003 \\ \times \quad 5 \\ \hline 5015 \end{array}$$

Divide and check.

1. $7\overline{)742}$
2. $6\overline{)612}$
3. $6\overline{)3047}$
4. $8\overline{)8720}$
5. $8\overline{)3254}$
6. $7\overline{)5643}$
7. $6\overline{)3664}$
8. $9\overline{)8910}$
9. $5\overline{)25{,}456}$
10. $7\overline{)56{,}756}$
11. $4\overline{)28{,}377}$
12. $8\overline{)73{,}674}$

PROBLEM SOLVING

13. A farm has 2525 hens separated equally in 5 henhouses. How many hens are in each henhouse?

____

14. An apple grower packs 816 select apples equally into 8 cases for shipment. How many apples will be in each case?

____

 Use with Lesson 3-5, text pages 104–105.

Short Division

Name ______________________

Date ______________________

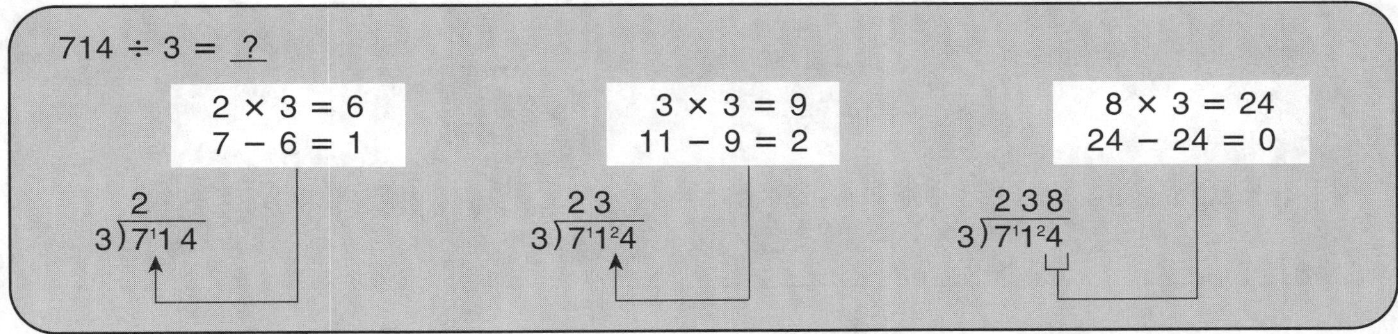

Use short division to divide. Then check.

1. 4)928	**2.** 6)150	**3.** 2)172	**4.** 5)75
5. 8)906	**6.** 7)2187	**7.** 3)843	**8.** 9)7668
9. 6)4637	**10.** 5)3317	**11.** 4)1937	**12.** 8)6103

Find the quotient. Use short division.

13. 5)325	**14.** 7)1830	**15.** 4)2576	**16.** 9)2905
17. 3)256	**18.** 6)478	**19.** 2)1350	**20.** 8)3539
21. 5)42,387	**22.** 7)434,714	**23.** 3)25,016	**24.** 9)302,485
25. 8)25,696	**26.** 6)308,850	**27.** 4)38,148	**28.** 5)393,455

PROBLEM SOLVING Use short division.

29. The total attendance at a 5-day fair was 631,760. If the attendance each day was the same, how many people visited the fair each day? ______________________

 Use with Lesson 3-6, text pages 106–107.

Divisibility and Mental Math

Name ______________________

Date ______________________

A number is divisible by:

2 if its ones digit is divisible by 2.

5 if its ones digit is 0 or 5.

10 if its ones digit is 0.

3 if the sum of its digits is divisible by 3.

9 if the sum of its digits is divisible by 9.

6 if it is divisible by both 2 and 3.

4 if its tens and ones digits form a number that is divisible by 4.

Write the numbers each is divisible by.

1. 30 ______

2. 48 ______

3. 252 ______

4. 453 ______

5. 404 ______

6. 625 ______

7. 1740 ______

8. 294 ______

9. 2367 ______

10. 3180 ______

11. 1830 ______

12. 3732 ______

13. 1455 ______

14. 9300 ______

15. 24,528 ______

16. 45,330 ______

17. 120,603 ______

18. 345,680 ______

PROBLEM SOLVING

19. A high school band of 100 musicians went on a trip to New York. Can the same number of musicians stay in three hotels? What might be a reasonable way to group the musicians? ______

20. A hotel manager needed to seat 250 guests at a wedding party. Can he seat them in groups of 5 or 10? ______

21. The manager has 122 flower arrangements. She wants to use all the arrangements and put the same number in each room. Should she put 2, 3, or 4 arrangements in each room? ______

Estimation: Compatible Numbers

Name ____________________

Date ____________________

Estimate:	Use compatible numbers:	Think:
3689 ÷ 7 = ?	3500 ÷ 7 = ?	3500 ÷ 7 = 500 So 3689 ÷ 7 is about 500.
41,345 ÷ 83 = ?	40,000 ÷ 80 = ?	40,000 ÷ 80 = 500 So 41,345 ÷ 83 is about 500.

Estimate each quotient. Use compatible numbers.

1. 459 ÷ 9 ________ **2.** 373 ÷ 4 ________ **3.** 5892 ÷ 8 ________ **4.** 2367 ÷ 7 ________

5. 921 ÷ 24 ________ **6.** 350 ÷ 23 ________ **7.** 375 ÷ 82 ________ **8.** 425 ÷ 58 ________

9. 325 ÷ 14 ________ **10.** 827 ÷ 43 ________ **11.** 888 ÷ 45 ________ **12.** 894 ÷ 29 ________

13. 1825 ÷ 32 ________ **14.** 3157 ÷ 44 ________ **15.** 1455 ÷ 53 ________ **16.** 2740 ÷ 84 ________

17. 97,617 ÷ 93 ________ **18.** 63,219 ÷ 78 ________ **19.** 54,413 ÷ 25 ________ **20.** 37,102 ÷ 57 ________

PROBLEM SOLVING

21. A company made 32,170 crayons. About how many boxes are needed if the crayons are packed in boxes of 16 each? ________

22. If 5816 passengers were carried by an airplane in a 23-day period and about the same number were carried each day, estimate the number of passengers carried each day. ________

23. In 32 hours, 83,697 cars crossed a bridge. About the same number crossed each hour. About how many cars crossed each hour? ________

24. Marvin has 3508 stamps about equally in 14 albums. About how many stamps are in each album? ________

25. A store has 963 books. About how many shelves are needed if each shelf holds 15 books? ________

Teens as Divisors

Name ______________________

Date ______________________

$12{,}416 \div 17 = \underline{?}$

Divide the hundreds.

$$\begin{array}{r} 8 \\ 17\overline{)12{,}416} \\ 136 \end{array}$$

Too large. Try 7.

$$\begin{array}{r} 7 \\ 17\overline{)12{,}416} \\ -119 \\ \hline 5 \end{array}$$

Divide the tens.

$$\begin{array}{r} 73 \\ 17\overline{)12{,}416} \\ -119\downarrow \\ \hline 51 \\ -51 \\ \hline 0 \end{array}$$

Divide the ones.

$$\begin{array}{r} 730 \text{ R6} \\ 17\overline{)12{,}416} \\ -119\downarrow \\ \hline 51 \\ -51\downarrow \\ \hline 06 \\ -0 \\ \hline 6 \end{array}$$

Find the quotient and the remainder. Then check.

1. $11\overline{)321}$ **2.** $13\overline{)435}$ **3.** $15\overline{)582}$ **4.** $17\overline{)6294}$

5. $19\overline{)7943}$ **6.** $12\overline{)8376}$ **7.** $14\overline{)92{,}462}$ **8.** $16\overline{)28{,}619}$

9. $18\overline{)47{,}829}$ **10.** $11\overline{)58{,}294}$ **11.** $15\overline{)183{,}247}$ **12.** $19\overline{)175{,}492}$

PROBLEM SOLVING

13. A bank received 116,714 key chains to give to customers. Each of the 14 tellers received the same number of key chains. At most, how many did each teller receive? How many chains were left over? ______________________

Two-Digit Divisors

Name ______________________

Date ______________________

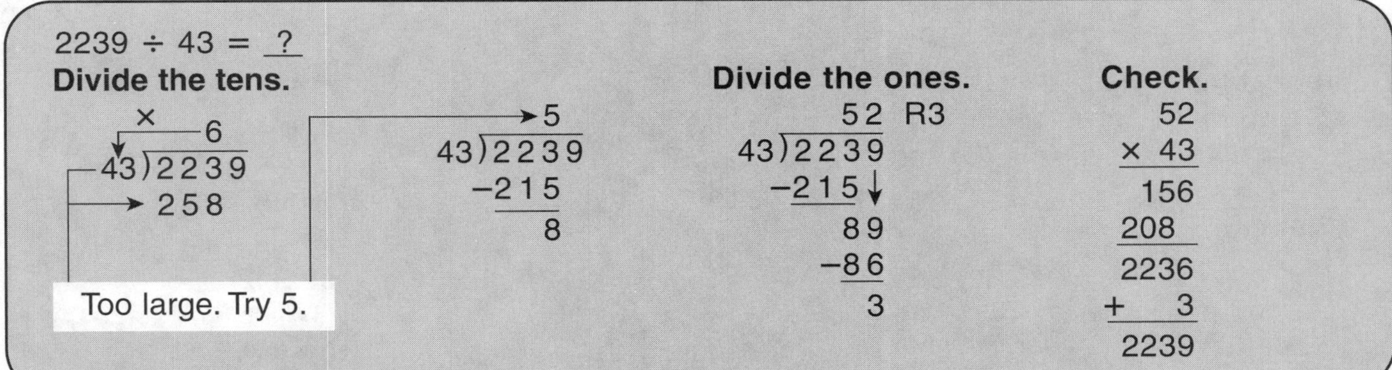

Divide and check.

1. 31)69 **2.** 82)4160 **3.** 46)1840 **4.** 17)360

5. 73)4526 **6.** 75)3718 **7.** 54)2870 **8.** 91)6643

9. 27)5807 **10.** 22)4500 **11.** 38)4294 **12.** 64)3399

PROBLEM SOLVING

13. A bird called the purple martin can eat 2400 mosquitos in 40 minutes. At that rate, how many can it eat per minute? ______________

14. At a picnic, 32 people ate 416 shrimp. Each person ate the same number of shrimp. How many shrimp did each person eat? ______________

15. Rachel took 188 pictures on her vacation. If each section of her photo album holds 24 pictures, how many sections could she fill? ______________

Dividing Larger Numbers

Name ______________________

Date ______________________

109,345 ÷ 23 = ?

Repeat the division steps until the division is completed.

Estimate.

$$23\overline{)109{,}345} \quad \text{estimate: } 4, \; 4 \times 23 = 92$$

Divide.

$$\begin{array}{r} 4{,}754 \text{ R3} \\ 23\overline{)109{,}345} \\ -92 \\ \hline 173 \\ -161 \\ \hline 124 \\ -115 \\ \hline 95 \\ -92 \\ \hline 3 \end{array}$$

Check.

$$\begin{array}{r} 4754 \\ \times \quad 23 \\ \hline 14\,262 \\ 95\,08 \\ \hline 109{,}342 \\ + \quad 3 \\ \hline 109{,}345 \end{array}$$

Divide and check.

1. $14\overline{)546}$
2. $18\overline{)5170}$
3. $19\overline{)17{,}784}$
4. $21\overline{)1379}$
5. $36\overline{)3312}$
6. $47\overline{)40{,}908}$
7. $42\overline{)10{,}083}$
8. $56\overline{)1344}$
9. $93\overline{)985{,}826}$
10. $33\overline{)178{,}829}$
11. $81\overline{)31{,}579}$

Use with Lesson 3-12, text pages 118–119.

Dividing Money

Name ____________________

Date ____________________

$1193.36 ÷ 14 = ?

To divide money:

- Place the dollar sign and the decimal point in the quotient.
- Divide as usual.

```
       $    .
14)$1193.36

       $  85.24
14)$1193.36
  -112
     73
    -70
      33
     -28
       56
      -56
        0
```

Divide and check.

1. 23)$17.25	**2.** 48)$41.28	**3.** 72)$136.08	**4.** 31)$17.98
5. 17)$15.13	**6.** 53)$35.51	**7.** 68)$621.52	**8.** 46)$162.84
9. 11)$94.93	**10.** 28)$26.32	**11.** 39)$1677.39	**12.** 17)$3485.34

PROBLEM SOLVING

13. The combined savings from summer jobs of 6 students was $825.78. How much did each student have if they all saved the same amount? ____________________

14. The cost of 14 identical sets of dishes is $1203.30. Find the cost of one set. ____________________

Order of Operations

Name ____________________

Date ____________________

Compute: $(80 \div 2) - 2 \times (3 \times 5) = \underline{\ ?\ }$

- Do the operations in parentheses first. — $(80 \div 2) - 2 \times (3 \times 5)$
- Multiply or divide from left to right. — $40 - 2 \times 15$
- Add or subtract from left to right. — $40 - 30$

$40 - 30 = 10$

Use the order of operations to compute.

1. $5 - 8 \div 2 + 7$ ______

2. $20 \div 4 + 3 \times 6$ ______

3. $(8 \times 7) + (56 \div 8)$ ______

4. $(42 - 12) \div (7 + 3)$ ______

5. $30 + 18 \div 3 - 12$ ______

6. $24 - 9 \div 3 \times 5$ ______

7. $2 + 6 \times 10 \div 30 + 7$ ______

8. $19 + 63 \div 9 \times 3 - 13$ ______

9. $59 - 35 \div 7 \times 4 + 53$ ______

10. $50 - 12 \div 3 \times 2$ ______

11. $20 \div 4 - 4 + (81 \div 9)$ ______

12. $25 - 6 \times 4 + (23 - 3) - 4$ ______

13. $(42 - 6) + 5 - 3 + (8 \times 3)$ ______

14. $3 + (37 - 1) \div 9 + (18 + 3)$ ______

15. $(5 \times 9) \div 5 + (8 \div 8)$ ______

16. $(64 \div 8) - 5 + (33 \times 3)$ ______

17. $(35 - 10) \div (4 + 1)$ ______

18. $20 + 6 \div (4 - 2)$ ______

19. $(6 \times 9) + (63 \div 7)$ ______

20. $32 \div 8 + 4 + (6 \times 0)$ ______

21. $(42 + 10) \div (4 \div 2)$ ______

22. $(55 - 15) \div (5 \times 2)$ ______

23. $(8 \times 8) + (9 \times 1)$ ______

24. $35 \div 7 + 5 - (7 \times 1)$ ______

Use with Lesson 3-14, text pages 122–123.

Problem-Solving Strategy: Make a Table/Find a Pattern

Name ____________________

Date ____________________

Wheels and Deals sells bicycles and tricycles. On their showroom floor, there are 4 times as many bicycles as tricycles. There is a total of 33 wheels on all of the cycles. How many of each kind of cycle are there?

Make a table to find different combinations.

There are 3 tricycles and 12 bicycles.

Tricycles	1	2	3
Bicycles	4	8	12
Tricycle Wheels	3	6	9
Bicycle Wheels	8	16	24
Total Wheels	11	22	33

Solve. Do your work on a separate sheet of paper.

1. There are 5 more crows than cows in a farmer's field. There is a total of 52 legs on all of the animals. How many crows and how many cows are in the field?

   ____

2. Hannah has the same number of nickels as Heather has dimes. Together, the coins are worth $1.65. How many coins does each girl have?

   ____

3. Bao and Calvin use 6 lemons to make every 4 quarts of lemonade. They want to make 12 quarts of lemonade. How many lemons do they need?

   ____

4. Dexter has 36 pets. He has 3 times as many fish as mice and half as many cats as mice. How many of each pet does Dexter have?

   ____

5. The temperature in Tampa was 55° F at 5 A.M. It rose 7° every hour for the next 2 hours and then rose 8° every hour for the next 3 hours after that. It then remained constant until 3 P.M. What was the temperature at noon?

   ____

6. Becky's dad is giving her money toward a new bicycle each time she washes the car. The first time he gave her $1.00. Each time after, he gave her $1.00 more than the time before. How much money will Becky have when she has washed the car 14 times?

   ____

7. Vonda has 3 more quarters than she has dimes in her pocket. She has no other coins. The total value of her coins is $1.80. How many quarters and how many dimes does Vonda have?

   ____

8. The temperature in Lakeview was 45° F at 7 A.M. It rose 5° each hour until noon. It remained steady for 3 hours. Then it fell 3° each hour until 8 P.M. What was the temperature in Lakeview at 8 P.M.?

   ____

Factors, Primes, and Composites

Name ____________________

Date ____________________

Factors are numbers that are multiplied to find a product.
Factors of 40: 1, 2, 4, 5, 8, 10, 20, 40

$1 \times 40 = 40$
$2 \times 20 = 40$
$4 \times 10 = 40$
$5 \times 8 = 40$

A prime number is greater than 1 and has exactly 2 factors, itself and 1.

A composite number is greater than 1 and has more than 2 factors.

List all the factors of each number.

1. 3 ________ **2.** 8 ________ **3.** 12 ________

4. 17 ________ **5.** 25 ________ **6.** 43 ________

7. 44 ________ **8.** 18 ________ **9.** 62 ________

10. 28 ________ **11.** 67 ________ **12.** 71 ________

13. 70 ________ **14.** 77 ________

List all the prime numbers in exercises 1–14.

15. ____________________

Complete each factor tree.

16.

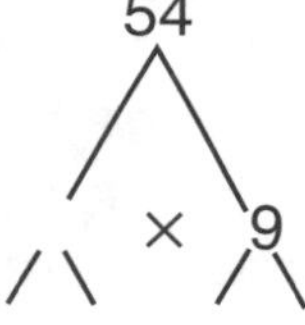

17.

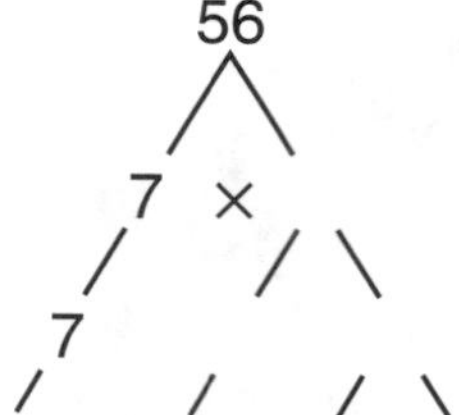

18.

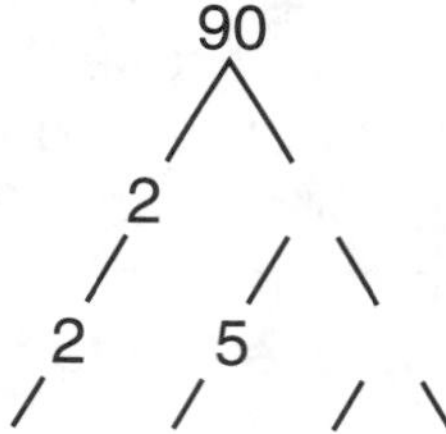

19.

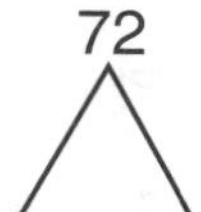

20.

21.

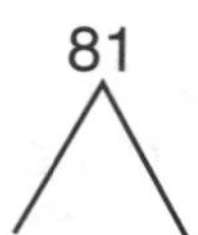

Greatest Common Factor

Name ______________________

Date ______________________

Greatest Common Factor of 6 and 15: ?
Factors of 6: **1**, 2, **3**, 6
Factors of 15: **1**, **3**, 5, 15
Common factors of 6 and 15: **1**, **3**
Greatest Common Factor (GCF): **3**

Complete the table.

	Number	Factors	Common Factors	GCF
1.	**8**			
	10			
2.	**12**			
	20			
3.	**6**			
	27			
4.	**15**			
	40			

List the common factors of each set of numbers. Then circle the GCF.

5. 9 and 12 ________ **6.** 27 and 18 ________ **7.** 10 and 20 ________

8. 14 and 12 ________ **9.** 36 and 20 ________ **10.** 42 and 56 ________

11. 8 and 16 ________ **12.** 30 and 18 ________ **13.** 11 and 33 ________

14. 13 and 24 ________ **15.** 6 and 24 ________ **16.** 5 and 10 ________

Find the GCF of each set of numbers.

17. 16 and 20 ____ **18.** 9 and 24 ____ **19.** 8 and 28 ____

20. 14 and 21 ____ **21.** 9 and 27 ____ **22.** 24 and 40 ____

23. 12 and 10 ____ **24.** 22 and 33 ____ **25.** 24 and 36 ____

PROBLEM SOLVING

26. What is the GCF of 13, 11, 24?

____

27. What is the GCF of 8, 12, 20?

____

Fraction Sense: Closer to 0, $\frac{1}{2}$, 1

Name ______________________

Date ______________________

Closer to 0	Closer to $\frac{1}{2}$	Closer to 1
Numerator is much less than denominator.	Numerator doubled is about equal to denominator.	Numerator is about equal to denominator.
$\frac{2}{10}$ $\frac{1}{6}$	$\frac{3}{7}$ $\frac{5}{9}$	$\frac{7}{8}$ $\frac{10}{12}$

Write whether each fraction is closer to *0*, $\frac{1}{2}$, or *1*.

1. $\frac{5}{8}$ ____ 2. $\frac{7}{12}$ ____ 3. $\frac{1}{3}$ ____ 4. $\frac{2}{9}$ ____ 5. $\frac{6}{7}$ ____ 6. $\frac{3}{5}$ ____

7. $\frac{2}{12}$ ____ 8. $\frac{8}{15}$ ____ 9. $\frac{17}{20}$ ____ 10. $\frac{9}{10}$ ____ 11. $\frac{5}{11}$ ____ 12. $\frac{3}{13}$ ____

13. $\frac{77}{100}$ ____ 14. $\frac{4}{25}$ ____ 15. $\frac{27}{50}$ ____ 16. $\frac{19}{100}$ ____ 17. $\frac{15}{18}$ ____ 18. $\frac{13}{50}$ ____

Estimate each fraction using compatible numbers.

19. $\frac{6}{21}$ ____ 20. $\frac{11}{32}$ ____ 21. $\frac{19}{39}$ ____ 22. $\frac{29}{32}$ ____ 23. $\frac{43}{81}$ ____ 24. $\frac{10}{99}$ ____

Complete.

25. $\frac{1 \times 4}{2 \times 4} = $ ____ 26. $\frac{6 \div 2}{8 \div 2} = $ ____ 27. $\frac{2 \times 3}{3 \times 3} = $ ____

28. $\frac{1}{4} = \frac{\quad}{8}$ 29. $\frac{8}{12} = \frac{\quad}{3}$ 30. $\frac{3}{4} = \frac{\quad}{20}$ 31. $\frac{10}{12} = \frac{\quad}{6}$

32. $\frac{12}{24} = \frac{1}{\quad}$ 33. $\frac{15}{25} = \frac{3}{\quad}$ 34. $\frac{3}{4} = \frac{18}{\quad}$ 35. $\frac{4}{5} = \frac{32}{\quad}$

36. $\frac{7}{50} = \frac{14}{\quad}$ 37. $\frac{21}{99} = \frac{\quad}{33}$ 38. $\frac{18}{72} = \frac{2}{\quad}$ 39. $\frac{1}{8} = \frac{\quad}{96}$

Write three equivalent fractions for each.

40. $\frac{8}{10}$ ______________ 41. $\frac{9}{12}$ ______________ 42. $\frac{10}{25}$ ______________

43. $\frac{6}{8}$ ______________ 44. $\frac{3}{5}$ ______________ 45. $\frac{1}{10}$ ______________

 Use with Lesson 4-4, text pages 140–141.

Fractions in Lowest Terms

Name ______________________

Date ______________________

Write as a fraction in lowest terms: $\frac{16}{18}$

Divide the numerator and the denominator by their GCF.

$\frac{16 \div 2}{18 \div 2} = \frac{8}{9}$ ⟵ **lowest terms**

Factors of 16: **1, 2,** 4, 8, 16

Factors of 18: **1, 2,** 3, 6, 9, 18

Common Factors: **1, 2**

Greatest Common Factor: **2**

Is each fraction in lowest terms? Write *Yes* or *No*.

1. $\frac{2}{10}$ ______ **2.** $\frac{8}{12}$ ______ **3.** $\frac{3}{10}$ ______ **4.** $\frac{7}{35}$ ______ **5.** $\frac{11}{33}$ ______

6. $\frac{5}{8}$ ______ **7.** $\frac{4}{18}$ ______ **8.** $\frac{6}{25}$ ______ **9.** $\frac{4}{9}$ ______ **10.** $\frac{6}{12}$ ______

11. $\frac{14}{26}$ ______ **12.** $\frac{18}{45}$ ______ **13.** $\frac{32}{44}$ ______ **14.** $\frac{18}{63}$ ______ **15.** $\frac{15}{23}$ ______

Write each fraction in simplest form.

16. $\frac{10}{25}$ ______ **17.** $\frac{8}{12}$ ______ **18.** $\frac{6}{16}$ ______ **19.** $\frac{9}{36}$ ______ **20.** $\frac{30}{50}$ ______

21. $\frac{14}{35}$ ______ **22.** $\frac{12}{16}$ ______ **23.** $\frac{2}{6}$ ______ **24.** $\frac{16}{18}$ ______ **25.** $\frac{2}{10}$ ______

26. $\frac{8}{40}$ ______ **27.** $\frac{3}{9}$ ______ **28.** $\frac{20}{60}$ ______ **29.** $\frac{6}{22}$ ______ **30.** $\frac{20}{24}$ ______

31. $\frac{9}{15}$ ______ **32.** $\frac{15}{20}$ ______ **33.** $\frac{10}{12}$ ______ **34.** $\frac{24}{32}$ ______ **35.** $\frac{21}{35}$ ______

36. $\frac{18}{36}$ ______ **37.** $\frac{27}{45}$ ______ **38.** $\frac{10}{25}$ ______ **39.** $\frac{54}{63}$ ______ **40.** $\frac{6}{30}$ ______

PROBLEM SOLVING Write each answer in lowest terms.

41. Carlene did 6 of her 10 homework examples correctly. What fractional part of her homework examples did she complete correctly? ______________

42. If Alberto used 9 of 36 nails, what fractional part of the nails did he use? ______________

43. There are 6 boys and 8 girls in a singing group. What fractional part of the group is girls? ______________

44. Four of 12 people chose milk. What fractional part of the group chose other drinks? ______________

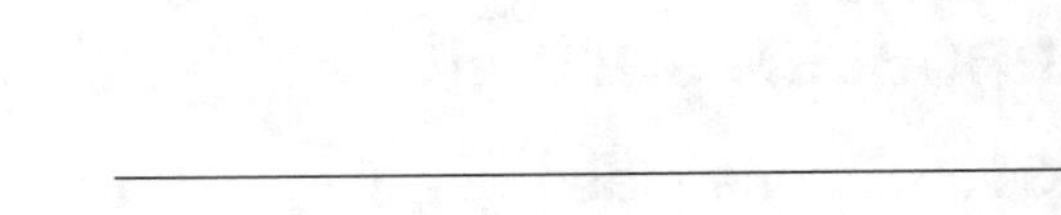

45. Jake had 15 sports cards. Five of the cards show baseball players. What fractional part of the cards show baseball players? ______________

Use with Lesson 4-5, text pages 142–143.

Fractions in Higher Terms

Name ____________________

Date ____________________

Rename a fraction as an equivalent fraction in higher terms by multiplying the numerator and denominator by the *same number.*

$\frac{3}{5} = \frac{3 \times 3}{5 \times 3} = \frac{9}{15}$ or $\frac{3}{5} = \frac{3 \times 6}{5 \times 6} = \frac{18}{30}$

Higher-Terms Fractions

Write the letter of the equivalent fraction in higher terms.

		a.	b.	c.	d.	e.
1. $\frac{1}{3}$	____	$\frac{5}{12}$	$\frac{3}{6}$	$\frac{4}{10}$	$\frac{5}{15}$	$\frac{9}{24}$
2. $\frac{2}{5}$	____	$\frac{8}{40}$	$\frac{5}{10}$	$\frac{8}{20}$	$\frac{6}{16}$	$\frac{9}{45}$
3. $\frac{7}{8}$	____	$\frac{15}{16}$	$\frac{28}{30}$	$\frac{35}{45}$	$\frac{48}{56}$	$\frac{56}{64}$
4. $\frac{1}{4}$	____	$\frac{5}{21}$	$\frac{8}{32}$	$\frac{6}{30}$	$\frac{2}{4}$	$\frac{3}{20}$
5. $\frac{2}{9}$	____	$\frac{14}{64}$	$\frac{8}{36}$	$\frac{8}{18}$	$\frac{12}{45}$	$\frac{16}{80}$
6. $\frac{1}{6}$	____	$\frac{2}{18}$	$\frac{6}{30}$	$\frac{18}{48}$	$\frac{9}{54}$	$\frac{6}{35}$

Write the missing term.

7. $\frac{3}{4} = \frac{9}{\_\_}$ **8.** $\frac{2}{3} = \frac{\_\_}{27}$ **9.** $\frac{1}{5} = \frac{\_\_}{10}$ **10.** $\frac{7}{8} = \frac{\_\_}{48}$ **11.** $\frac{2}{9} = \frac{\_\_}{18}$

12. $\frac{4}{7} = \frac{\_\_}{35}$ **13.** $\frac{1}{3} = \frac{6}{\_\_}$ **14.** $\frac{2}{9} = \frac{\_\_}{36}$ **15.** $\frac{3}{8} = \frac{\_\_}{24}$ **16.** $\frac{1}{2} = \frac{5}{\_\_}$

17. $\frac{3}{7} = \frac{\_\_}{21}$ **18.** $\frac{5}{6} = \frac{\_\_}{24}$ **19.** $\frac{2}{3} = \frac{6}{\_\_}$ **20.** $\frac{1}{6} = \frac{4}{\_\_}$ **21.** $\frac{4}{9} = \frac{\_\_}{45}$

Write the equivalent fractions.

22. $\frac{3}{4} = \frac{\_\_}{16}$ **23.** $\frac{1}{8} = \frac{3}{\_\_}$ **24.** $\frac{2}{5} = \frac{\_\_}{15}$ **25.** $\frac{9}{10} = \frac{27}{\_\_}$ **26.** $\frac{5}{12} = \frac{15}{\_\_}$

27. $\frac{2}{7} = \frac{44}{\_\_} = \frac{\_\_}{28} = \frac{\_\_}{42} = \frac{14}{\_\_} = \frac{18}{\_\_}$ **28.** $\frac{4}{9} = \frac{8}{\_\_} = \frac{16}{\_\_} = \frac{\_\_}{54} = \frac{32}{\_\_} = \frac{36}{\_\_}$

29. $\frac{1}{5} = \frac{\_\_}{15} = \frac{5}{\_\_} = \frac{\_\_}{35} = \frac{\_\_}{45} = \frac{11}{\_\_}$ **30.** $\frac{3}{11} = \frac{\_\_}{22} = \frac{\_\_}{33} = \frac{12}{\_\_} = \frac{\_\_}{55} = \frac{21}{\_\_}$

PROBLEM SOLVING

31. Suzy's class used $\frac{5}{8}$ of the cafeteria trays for their science display. There are 72 trays in all. Write a fraction in higher terms that shows how many trays were used. ____________________

 Use with Lesson 4-6, text pages 144–145.

Multiples: LCM and LCD

Name ______________________

Date ______________________

Multiples of 4: 0, 4, 8, **12,** 16, 20, **24,** 28, . . .
Multiples of 6: 0, 6, **12,** 18, **24,** 30, 36, 42, . . .
Nonzero common multiples of 4 and 6: **12, 24,** . . .
Least Common Multiple (LCM) of 4 and 6: **12**
Least Common Denominator (LCD) of $\frac{1}{4}$ and $\frac{1}{6}$: **12**

List the first ten nonzero multiples of each.

1. 9 ______________________
2. 7 ______________________
3. 5 ______________________
4. 8 ______________________

List the first four common multiples of each set of numbers.

5. 2, 4 ____________
6. 2, 3 ____________
7. 3, 5 ____________
8. 4, 3 ____________
9. 6, 8 ____________
10. 5, 10 ____________
11. 4, 8 ____________
12. 6, 12 ____________
13. 2, 3, 4 ____________
14. 4, 6, 8 ____________

Write the least common multiple (LCM) of each set of numbers.

15. 4 and 10 ____
16. 3 and 5 ____
17. 6 and 8 ____
18. 6 and 12 ____
19. 7 and 8 ____
20. 5 and 8 ____
21. 2 and 6 ____
22. 4 and 5 ____
23. 4 and 16 ____
24. 4 and 12 ____
25. 6 and 10 ____
26. 4 and 8 ____
27. 2 and 4 ____
28. 5 and 7 ____
29. 7 and 9 ____
30. 3 and 7 ____
31. 2 and 3 ____
32. 5 and 6 ____
33. 5 and 20 ____
34. 5 and 2 ____
35. 9 and 4 ____
36. 7 and 2 ____
37. 2 and 10 ____
38. 3 and 8 ____

Write the least common denominator (LCD) of each set of fractions.

39. $\frac{2}{8}$ and $\frac{3}{12}$ ____
40. $\frac{1}{3}$ and $\frac{4}{5}$ ____
41. $\frac{5}{16}$ and $\frac{1}{4}$ ____
42. $\frac{2}{9}$ and $\frac{3}{4}$ ____
43. $\frac{5}{6}$ and $\frac{1}{8}$ ____
44. $\frac{3}{8}$ and $\frac{11}{16}$ ____
45. $\frac{2}{3}$ and $\frac{5}{6}$ ____
46. $\frac{1}{4}$ and $\frac{1}{3}$ ____
47. $\frac{1}{2}$ and $\frac{4}{5}$ ____
48. $\frac{6}{7}$ and $\frac{2}{3}$ ____
49. $\frac{3}{4}$ and $\frac{1}{6}$ ____
50. $\frac{1}{4}$ and $\frac{7}{8}$ ____

Mixed Numbers and Improper Fractions

Name ____________________

Date ____________________

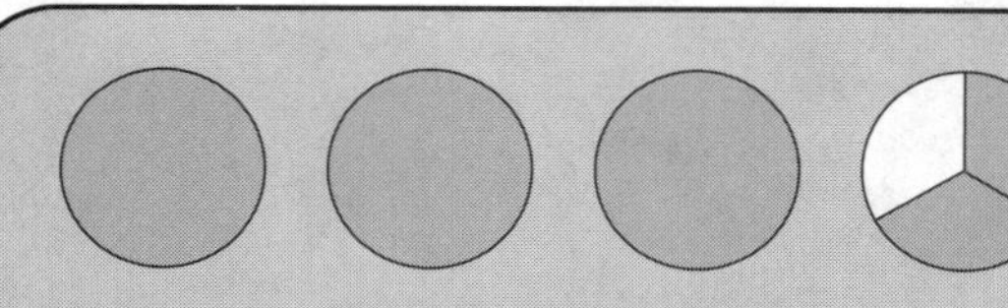

Read: three and two thirds

Write: $3\frac{2}{3}$

Rename $\frac{28}{8}$ as a whole number or a mixed number in simplest form.

$\frac{28}{8} = 8\overline{)28}$ → 3 R4 $= 3\frac{4}{8} = 3\frac{1}{2}$

Write a mixed number that represents the shaded part.

1.   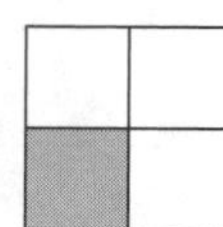 ______

2. 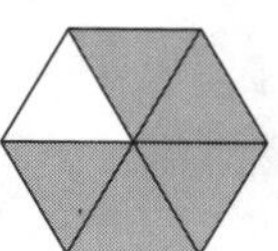 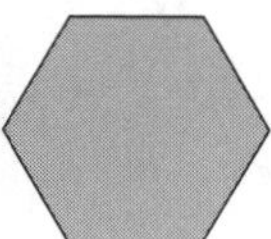 ______

Write as a mixed number.

3. five and three tenths ______ **4.** three and seven twelfths ______

5. eight and one fifth ______ **6.** two and nine elevenths ______

7. ten and six sevenths ______ **8.** fifteen and two thirds ______

Round each to the nearest whole number.

9. $7\frac{9}{10}$ ____ **10.** $4\frac{3}{4}$ ____ **11.** $6\frac{1}{2}$ ____ **12.** $8\frac{1}{3}$ ____ **13.** $9\frac{3}{8}$ ____

Write a numerator to give each a value equal to 1.

14. $\frac{\_\_}{14}$ **15.** $\frac{\_\_}{20}$ **16.** $\frac{\_\_}{12}$ **17.** $\frac{\_\_}{21}$ **18.** $\frac{\_\_}{17}$ **19.** $\frac{\_\_}{19}$ **20.** $\frac{\_\_}{16}$

Write a numerator to give each a value greater than 1.

21. $\frac{\_\_}{3}$ **22.** $\frac{\_\_}{5}$ **23.** $\frac{\_\_}{6}$ **24.** $\frac{\_\_}{12}$ **25.** $\frac{\_\_}{9}$ **26.** $\frac{\_\_}{15}$ **27.** $\frac{\_\_}{20}$

Write a whole number or a mixed number in simplest form.

28. $\frac{17}{5} =$ ____ **29.** $\frac{20}{7} =$ ____ **30.** $\frac{11}{6} =$ ____ **31.** $\frac{20}{5} =$ ____ **32.** $\frac{14}{3} =$ ____

33. $\frac{54}{6} =$ ____ **34.** $\frac{16}{3} =$ ____ **35.** $\frac{36}{4} =$ ____ **36.** $\frac{37}{7} =$ ____ **37.** $\frac{7}{4} =$ ____

38. $\frac{10}{9} =$ ____ **39.** $\frac{21}{3} =$ ____ **40.** $\frac{44}{7} =$ ____ **41.** $\frac{21}{8} =$ ____ **42.** $\frac{38}{5} =$ ____

43. $\frac{12}{2} =$ ____ **44.** $\frac{16}{16} =$ ____ **45.** $\frac{65}{8} =$ ____ **46.** $\frac{42}{10} =$ ____ **47.** $\frac{8}{8} =$ ____

Comparing and Ordering Fractions

Name ______________________

Date ______________________

Compare: $\frac{2}{5}$ ___?___ $\frac{1}{3}$

LCD: 15

$\frac{2 \times 3}{5 \times 3} = \frac{6}{15}$ $\frac{1 \times 5}{3 \times 5} = \frac{5}{15}$

$6 > 5 \rightarrow \frac{6}{15} > \frac{5}{15}$

So $\frac{2}{5} > \frac{1}{3}$.

Order from least to greatest: $\frac{4}{5}$, $\frac{1}{4}$, $\frac{3}{10}$

LCD: 20

$\frac{4}{5} = \frac{16}{20}$; $\frac{1}{4} = \frac{5}{20}$; $\frac{3}{10} = \frac{6}{20}$

$\frac{5}{20} < \frac{6}{20} < \frac{16}{20}$ Think: $5 < 6 < 16$

Least to greatest: $\frac{1}{4}$, $\frac{3}{10}$, $\frac{4}{5}$

Compare. Write <, =, or >.

1. $\frac{3}{10}$ ____ $\frac{5}{20}$
2. $\frac{4}{32}$ ____ $\frac{7}{35}$
3. $\frac{7}{12}$ ____ $\frac{7}{24}$
4. $\frac{1}{9}$ ____ $\frac{1}{8}$
5. $\frac{2}{6}$ ____ $\frac{3}{8}$
6. $\frac{4}{7}$ ____ $\frac{2}{3}$
7. $\frac{5}{10}$ ____ $\frac{1}{2}$
8. $\frac{6}{8}$ ____ $\frac{3}{4}$
9. $\frac{3}{8}$ ____ $\frac{2}{5}$
10. $\frac{7}{10}$ ____ $\frac{2}{3}$
11. $\frac{2}{3}$ ____ $\frac{1}{4}$
12. $\frac{8}{20}$ ____ $\frac{2}{5}$
13. $\frac{4}{5}$ ____ $\frac{9}{12}$
14. $\frac{8}{12}$ ____ $\frac{4}{6}$
15. $\frac{9}{20}$ ____ $\frac{1}{4}$
16. $\frac{3}{8}$ ____ $\frac{1}{2}$
17. $\frac{15}{12}$ ____ $\frac{20}{16}$
18. $2\frac{1}{5}$ ____ $2\frac{3}{10}$
19. $4\frac{3}{8}$ ____ 4
20. 2 ____ $\frac{12}{7}$

Write in order from least to greatest.

21. $\frac{3}{8}$, $\frac{1}{7}$, $\frac{1}{2}$ ______________
22. $\frac{2}{3}$, $\frac{1}{3}$, $\frac{7}{10}$ ______________
23. $3\frac{5}{8}$, $3\frac{1}{2}$, $3\frac{3}{8}$ ______________
24. $12\frac{1}{8}$, $12\frac{3}{10}$, $12\frac{1}{4}$ ______________
25. $\frac{2}{3}$, $\frac{3}{4}$, $\frac{1}{6}$ ______________
26. $2\frac{7}{8}$, $2\frac{1}{2}$, $2\frac{1}{3}$ ______________

Write in order from greatest to least.

27. $\frac{1}{3}$, $\frac{1}{5}$, $\frac{4}{15}$ ______________
28. $\frac{1}{8}$, $\frac{9}{16}$, $\frac{1}{4}$ ______________
29. $\frac{1}{12}$, $\frac{3}{4}$, $\frac{3}{12}$ ______________
30. $\frac{7}{4}$, $1\frac{1}{3}$, $1\frac{4}{9}$ ______________
31. $\frac{1}{4}$, $\frac{1}{5}$, $\frac{3}{10}$ ______________
32. $5\frac{5}{6}$, $5\frac{3}{4}$, $5\frac{11}{12}$ ______________

PROBLEM SOLVING

33. Adele and Morris took turns driving their car across the country. Adele drove $\frac{2}{3}$ of the way, and Morris drove $\frac{4}{12}$ of the way. Who drove less? ______________

Problem-Solving Strategy: Organized List

Name ______________________________

Date ______________________________

Amos, Beth, Carl, and Dawn want to ride on the Ferris wheel. There is one empty car and only 2 people can ride in a car. In how many different ways can 2 friends ride in the empty car?

Make an organized list of the possible pairs. Count the number of pairs.

There are 6 ways that 2 friends can ride in the empty car.

Amos ⟷ Beth
Amos ⟷ Carl
Amos ⟷ Dawn
Beth ⟷ Carl
Beth ⟷ Dawn
Carl ⟷ Dawn

Solve. Do your work on a separate sheet of paper.

1. Nico, Paul, and Mary line up single file. How many different ways can they line up?

____

2. Hidori, Cherie, Ellis, and Morgan are running a race. In how many different ways might they finish the race?

____

3. You have a pocket full of nickels, dimes, and quarters to make a call. The call costs 75¢. How many different combinations of coins can you use to pay for the call?

____

4. Natasha has a red sweater and a white sweater. She can wear them with a blue skirt, yellow skirt, or black skirt. How many different outfits can she make?

____

5. You are at a party where crackers and cheese spreads are among the refreshments. You can choose any one of 4 kinds of spreads and put on it any one of 3 kinds of crackers. How many different choices do you have?

____

6. Julio, Barry, Lianne, Carl, Dennis, and Elena visit the amusement park. They want to ride on the bumper cars. Each car can hold two people. How many different ways can the friends ride together in a car?

____

7. Oscar, Max, Norma, Pasha, and Hank want to ride. There are 4 horses — Spanky, Rita, Buttercup, and Ebony. In how many different ways can the horses be assigned if Max and Oscar ride on Spanky?

____

8. A building has 2 front doors, 2 doors on the left side, 2 back doors and 2 doors on the right side. How many different ways can you enter and leave the building if you enter through a front door and then leave through any of the other doors?

____

Adding Fractions: Unlike Denominators

Name ______________________

Date ______________________

Add: $\frac{1}{8} + \frac{1}{3} = \underline{?}$

Think: LCD is 24.

Rename the fractions. Then add.

$$\begin{array}{rcccc} & \frac{1}{8} & = & \frac{1 \times 3}{8 \times 3} & = & \frac{3}{24} \\ + & \frac{1}{3} & = & \frac{1 \times 8}{3 \times 8} & = & \frac{8}{24} \\ \hline & & & & & \frac{11}{24} \end{array}$$

$\frac{11}{24}$ ← simplest form

Complete. Then add. Write the sum in simplest form.

1. $\frac{5}{6} = \frac{20}{}$; $+\frac{1}{8} = +\frac{}{24}$

2. $1\frac{3}{4} = \frac{15}{}$; $+\frac{1}{5} = +\frac{}{20}$

3. $\frac{1}{4} = \frac{}{16}$; $+\frac{7}{16} = +\frac{}{16}$

4. $\frac{1}{2} = \frac{3}{}$; $+\frac{1}{3} = +\frac{}{6}$

Add. Write each sum in simplest form.

5. $\frac{2}{3} + \frac{2}{15}$

6. $\frac{2}{3} + \frac{1}{7}$

7. $\frac{4}{9} + \frac{3}{6}$

8. $\frac{1}{4} + \frac{1}{5}$

9. $\frac{3}{10} + \frac{1}{2}$

10. $\frac{3}{4} + \frac{5}{12}$

11. $\frac{2}{3} + \frac{4}{5}$

12. $\frac{5}{7} + \frac{3}{4}$

13. $\frac{3}{5} + \frac{7}{10}$

14. $\frac{8}{15} + \frac{5}{6}$

15. $\frac{1}{6} + \frac{1}{4} =$ ______

16. $\frac{1}{4} + \frac{5}{8} =$ ______

17. $\frac{2}{3} + \frac{1}{18} =$ ______

18. $\frac{4}{9} + \frac{2}{5} =$ ______

19. $\frac{3}{8} + \frac{2}{7} =$ ______

20. $\frac{3}{4} + \frac{1}{8} =$ ______

PROBLEM SOLVING

21. Helena walks $\frac{5}{8}$ mi to her friend's house and then $\frac{1}{4}$ mi to school. How far does she walk? ______________________

22. What is the sum of three fourths and five eighths? ______________________

Adding Three Fractions

Name ______________________

Date ______________________

Add: $\frac{2}{3} + \frac{5}{6} + \frac{4}{9} = \underline{?}$

Think: LCD is 18.

$$\frac{2}{3} = \frac{2 \times 6}{3 \times 6} = \frac{12}{18}$$
$$\frac{5}{6} = \frac{5 \times 3}{6 \times 3} = \frac{15}{18}$$
$$+\frac{4}{9} = \frac{4 \times 2}{9 \times 2} = \frac{8}{18}$$
$$\frac{35}{18} = 1\frac{17}{18}$$

Add. Write each sum in simplest form.

1. $\frac{1}{7} + \frac{3}{7} + \frac{1}{7}$

2. $\frac{2}{11} + \frac{4}{11} + \frac{3}{11}$

3. $\frac{6}{14} + \frac{4}{14} + \frac{4}{14}$

4. $\frac{2}{15} + \frac{3}{15} + \frac{12}{15}$

5. $\frac{5}{17} + \frac{4}{17} + \frac{7}{17}$

6. $\frac{3}{12} + \frac{2}{3} + \frac{1}{4}$

7. $\frac{1}{2} + \frac{1}{10} + \frac{1}{5}$

8. $\frac{2}{9} + \frac{1}{6} + \frac{4}{18}$

9. $\frac{1}{3} + \frac{3}{7} + \frac{4}{21}$

10. $\frac{1}{4} + \frac{3}{16} + \frac{2}{8}$

11. $\frac{3}{10} + \frac{1}{5} + \frac{4}{15} =$ ________

12. $\frac{3}{4} + \frac{7}{12} + \frac{1}{6} =$ ________

13. $\frac{2}{9} + \frac{3}{9} + \frac{1}{3} =$ ________

14. $\frac{3}{4} + \frac{1}{2} + \frac{3}{8} =$ ________

15. $\frac{1}{5} + \frac{1}{4} + \frac{7}{10} =$ ________

16. $\frac{7}{18} + \frac{1}{3} + \frac{5}{6} =$ ________

PROBLEM SOLVING

17. A family used $\frac{5}{8}$, $\frac{3}{4}$, and $\frac{1}{2}$ gallons of water on 3 different days. How many gallons of water did the family use in all? ________

18. Deirdre exercised for $\frac{3}{4}$ h on Thursday, $\frac{1}{2}$ h on Friday, and $\frac{11}{12}$ h on Monday. How many hours did she exercise in all? ________

Adding Mixed Numbers

Name ______________________

Date ______________________

Find the LCD. Add fractions.	Add whole numbers.	Write the sum in simplest form.
$3\frac{1}{5} = 3\frac{2}{10}$	$3\frac{1}{5}$	$3\frac{1}{5}$
$+5\frac{3}{10} = +5\frac{3}{10}$	$+5\frac{3}{10}$	$+5\frac{3}{10}$
$\frac{5}{10}$	$8\frac{5}{10}$	$8\frac{1}{2}$

Add. Write each sum in simplest form.

1. $2\frac{2}{5} + 7\frac{1}{5}$ **2.** $5\frac{2}{15} + 7\frac{1}{15}$ **3.** $3\frac{2}{9} + 4\frac{4}{9}$ **4.** $\frac{1}{3} + 2\frac{1}{3}$ **5.** $14\frac{7}{18} + 9\frac{1}{18}$ **6.** $4\frac{8}{12} + 2\frac{3}{12}$

7. $1\frac{1}{4} + 3\frac{1}{2}$ **8.** $4\frac{2}{12} + 3\frac{1}{4}$ **9.** $5\frac{1}{3} + 3\frac{1}{2}$ **10.** $7\frac{1}{6} + \frac{1}{3}$ **11.** $8\frac{1}{3} + 5\frac{1}{12}$ **12.** $9\frac{1}{7} + 1\frac{2}{21}$

13. $7\frac{3}{5} + 5\frac{2}{15} =$ ____ **14.** $9\frac{2}{5} + 2\frac{7}{20} =$ ____ **15.** $4\frac{1}{3} + 1\frac{2}{9} + 1\frac{5}{18} =$ ____

16. $6\frac{1}{5} + 5\frac{4}{6} =$ ____ **17.** $\frac{1}{5} + 3\frac{2}{10} =$ ____ **18.** $4\frac{2}{5} + 2\frac{1}{4} + 3\frac{1}{20} =$ ____

19. $3\frac{3}{4} + 8\frac{1}{16} =$ ____ **20.** $5\frac{1}{4} + 7\frac{3}{8} =$ ____ **21.** $7\frac{2}{15} + 3\frac{3}{10} + 1\frac{1}{5} =$ ____

PROBLEM SOLVING

22. Amiel weighs $12\frac{1}{2}$ pounds and Marthe weighs $21\frac{1}{4}$ pounds. What is the total weight of both babies? ______________

23. Nan bought $6\frac{1}{8}$ pounds of fish. Hank bought $3\frac{1}{2}$ pounds of fish. How much fish did the two buy altogether? ______________

24. Ms. Johnson bought $2\frac{3}{8}$ yd of fabric to make a dress and $1\frac{1}{3}$ yd of fabric to make a shirt. How much fabric did she buy? ______________

Renaming Mixed-Number Sums

Name ______________________

Date ______________________

$8\frac{2}{3} + 4\frac{7}{9} = \underline{?}$

$$\begin{array}{rcl} 8\frac{2}{3} & = & 8\frac{6}{9} \\ +\,4\frac{7}{9} & = & +\,4\frac{7}{9} \\ \hline & & 12\frac{13}{9} = 12 + 1\frac{4}{9} = 13\frac{4}{9} \end{array}$$

$1\frac{5}{12} + 2\frac{1}{4} + 1\frac{1}{3} = \underline{?}$

$$\begin{array}{rcl} 1\frac{5}{12} & = & 1\frac{5}{12} \\ 2\frac{1}{4} & = & 2\frac{3}{12} \\ +\,1\frac{1}{3} & = & +\,1\frac{4}{12} \\ \hline & & 4\frac{12}{12} = 4 + 1 = 5 \end{array}$$

Rename each as a whole number or a mixed number in simplest form.

1. $\frac{5}{3}$ ______ **2.** $6\frac{2}{2}$ ______ **3.** $5\frac{6}{4}$ ______ **4.** $6\frac{7}{4}$ ______ **5.** $8\frac{12}{10}$ ______

6. $7\frac{18}{14}$ ______ **7.** $2\frac{16}{15}$ ______ **8.** $5\frac{15}{11}$ ______ **9.** $7\frac{20}{15}$ ______ **10.** $42\frac{20}{20}$ ______

Add.

11. $8\frac{2}{3} + 3\frac{2}{3}$

12. $6\frac{4}{5} + 2\frac{3}{5}$

13. $4\frac{5}{8} + 7\frac{3}{8}$

14. $4\frac{7}{24} + 7\frac{5}{6}$

15. $9\frac{3}{4} + 5\frac{1}{4}$

16. $3\frac{3}{5} + 3\frac{5}{12}$

17. $2\frac{2}{15} + 3\frac{3}{15}$

18. $1\frac{4}{7} + 5\frac{1}{2}$

19. $13\frac{11}{12} + 4\frac{5}{6}$

20. $12\frac{7}{18} + 3\frac{7}{9}$

21. $2\frac{5}{8}$, $4\frac{1}{2}$, $+\,3\frac{3}{16}$

22. $4\frac{1}{3}$, $6\frac{5}{12}$, $+\,1\frac{1}{4}$

23. $3\frac{3}{20}$, $4\frac{1}{5}$, $+\,3\frac{3}{4}$

24. $5\frac{2}{7}$, $2\frac{4}{14}$, $+\,\frac{5}{7}$

25. $1\frac{5}{15}$, $8\frac{1}{10}$, $+\,6\frac{5}{6}$

26. $1\frac{3}{8} + 6\frac{2}{3} + 4\frac{7}{24} =$ ______

27. $3\frac{2}{5} + 2\frac{7}{15} + 5\frac{3}{10} =$ ______

PROBLEM SOLVING

28. Charlene mixed $2\frac{5}{8}$ gal of lemonade with $3\frac{3}{4}$ gal of seltzer to make punch. How much punch did she make? ______

 Use with Lesson 5-5, text pages 172–173.

Subtracting: Unlike Denominators

Name ____________________

Date ____________________

Subtract: $\frac{7}{10} - \frac{1}{5} = \underline{?}$

Think: LCD is 10.

Rename the fractions.
Then subtract.

$$\begin{array}{rcccl} \frac{7}{10} & & & = & \frac{7}{10} \\ -\frac{1}{5} & = & \frac{1 \times 2}{5 \times 2} & = & \frac{2}{10} \\ \hline & & & & \frac{5}{10} = \frac{1}{2} \end{array}$$

← lowest terms

Subtract. Write the difference in lowest terms.

1. $\frac{2}{3} - \frac{7}{12}$
2. $\frac{3}{4} - \frac{1}{2}$
3. $\frac{3}{8} - \frac{1}{4}$
4. $\frac{7}{9} - \frac{2}{3}$
5. $\frac{3}{8} - \frac{3}{16}$
6. $\frac{9}{10} - \frac{4}{5}$
7. $\frac{7}{10} - \frac{2}{5}$
8. $\frac{8}{9} - \frac{1}{3}$
9. $\frac{5}{8} - \frac{1}{4}$
10. $\frac{7}{9} - \frac{2}{3}$
11. $\frac{11}{12} - \frac{5}{6}$
12. $\frac{5}{6} - \frac{1}{3}$
13. $\frac{7}{16} - \frac{1}{4}$
14. $\frac{8}{15} - \frac{1}{3}$
15. $\frac{9}{10} - \frac{1}{2}$
16. $\frac{7}{30} - \frac{1}{10} =$ ____________________
17. $\frac{5}{12} - \frac{3}{24} =$ ____________________
18. $\frac{10}{16} - \frac{3}{8} =$ ____________________
19. $\frac{5}{8} - \frac{3}{40} =$ ____________________

PROBLEM SOLVING

20. Josefina spent $\frac{1}{4}$ hour exercising and another $\frac{3}{20}$ hour shooting baskets. How much more time did she spend exercising? ____________________

21. Gareth ate $\frac{3}{8}$ c of soup. Dan ate $\frac{1}{4}$ c of soup. Who ate more soup? How much more? ____________________

Use with Lesson 5-7, text pages 176–177.

More Subtraction of Fractions

Name ______________________

Date ______________________

Subtract: $\frac{5}{5} - \frac{6}{8} = \underline{\ ?\ }$

Think: LCD is 24.

Rename. Then subtract.

$$\begin{array}{rcccc} \frac{5}{6} & = & \frac{5 \times 4}{6 \times 4} & = & \frac{20}{24} \\ -\frac{6}{8} & = & \frac{6 \times 3}{8 \times 3} & = & \frac{18}{24} \\ \hline & & & & \frac{2}{24} = \frac{1}{12} \end{array}$$

← simplest form

Subtract.

1. $\frac{1}{6} - \frac{1}{10}$

2. $\frac{7}{10} - \frac{5}{12}$

3. $\frac{5}{6} - \frac{1}{8}$

4. $\frac{5}{8} - \frac{1}{20}$

5. $\frac{5}{8} - \frac{1}{3}$

6. $\frac{4}{5} - \frac{1}{2}$

7. $\frac{6}{11} - \frac{1}{2}$

8. $\frac{1}{12} - \frac{1}{15}$

9. $\frac{6}{7} - \frac{1}{3}$

10. $\frac{6}{10} - \frac{1}{4}$

11. $\frac{3}{4} - \frac{1}{3} =$ ______________

12. $\frac{3}{8} - \frac{1}{5} =$ ______________

13. $\frac{1}{2} - \frac{1}{7} =$ ______________

14. $\frac{9}{10} - \frac{3}{4} =$ ______________

Compare. Write <, =, or >.

15. $\frac{1}{2} - \frac{1}{3}$ __________ $\frac{1}{6} + \frac{1}{6}$

16. $\frac{1}{2} - \frac{1}{9}$ __________ $\frac{1}{3} + \frac{1}{18}$

PROBLEM SOLVING

17. Esther had $\frac{7}{8}$ yd of ribbon. She cut off $\frac{1}{3}$ yd. How much ribbon did she have left? ______________

18. In Guatemala, $\frac{6}{10}$ of an inch of rain fell in one week. In Honduras, $\frac{5}{8}$ of an inch fell in the same time. Where did more rain fall? How much more? ______________

Use with Lesson 5-8, text pages 178–179.

Subtracting Mixed Numbers

Name ______________________

Date ______________________

$\begin{array}{r} 9\frac{5}{8} \\ -\,2\frac{1}{8} \\ \hline 7\frac{4}{8} = 7\frac{1}{2} \end{array}$ Remember: Subtract fractions. Then subtract whole numbers.	$\begin{array}{rcl} 8\frac{5}{6} & = & 8\frac{20}{24} \\ -2\frac{2}{8} & = & -\,2\frac{6}{24} \\ \hline & & 6\frac{14}{24} = 6\frac{7}{12} \end{array}$ Think: LCD is 24.

Subtract.

1. $4\frac{7}{9} - \frac{2}{5}$
2. $6\frac{8}{9} - 2\frac{2}{3}$
3. $7\frac{11}{12} - 5\frac{3}{4}$
4. $4\frac{7}{8} - 3\frac{1}{2}$
5. $9\frac{2}{5} - 4\frac{1}{20}$
6. $3\frac{13}{15} - 3\frac{5}{6}$
7. $7\frac{8}{9} - 5\frac{3}{4}$
8. $5\frac{9}{10} - 2\frac{3}{4}$
9. $8\frac{4}{7} - 8\frac{1}{3}$
10. $7\frac{5}{6} - 4\frac{2}{5}$
11. $8\frac{7}{8} - 3\frac{2}{7}$
12. $7\frac{4}{5} - 2\frac{1}{2}$
13. $5\frac{6}{12} - 1\frac{3}{7}$
14. $2\frac{9}{10} - 2\frac{1}{5}$
15. $4\frac{4}{15} - 3\frac{3}{30}$

Circle the letter of the correct answer.

16. $6\frac{7}{8} - 2\frac{7}{8}$	**a.** $4\frac{7}{8}$	**b.** 8	**c.** 4	**d.** 5
17. $8\frac{2}{3} - 1\frac{1}{15}$	**a.** $7\frac{8}{15}$	**b.** $7\frac{4}{5}$	**c.** $7\frac{1}{12}$	**d.** $7\frac{3}{5}$
18. $5\frac{2}{5} - 2\frac{1}{8}$	**a.** $3\frac{1}{3}$	**b.** $3\frac{1}{40}$	**c.** $3\frac{11}{40}$	**d.** $3\frac{1}{4}$
19. $5\frac{4}{7} - 4\frac{4}{7}$	**a.** $2\frac{1}{2}$	**b.** $2\frac{1}{7}$	**c.** $1\frac{2}{7}$	**d.** 1

PROBLEM SOLVING

20. Clarissa bought $2\frac{2}{3}$ yd of wrapping paper. After wrapping a gift, she had $1\frac{1}{6}$ yd left. How much did she use? ______________

21. Muni painted $4\frac{1}{2}$ feet of a $10\frac{3}{4}$-ft long fence. How much of the fence was not painted? ______________

 Use with Lesson 5-9, text pages 180–181.

Subtraction With Renaming

Name ______________________

Date ______________________

Subtract: $8 - 3\frac{6}{8} = \underline{\ ?\ }$

$$\begin{array}{rl} 8 & = 7\frac{8}{8} \\ -3\frac{6}{8} & = 3\frac{6}{8} \\ \hline & 4\frac{2}{8} = 4\frac{1}{4} \end{array}$$

$8 = 7 + 1$

$= 7 + \frac{8}{8}$

$= 7\frac{8}{8}$

Simplest form

Complete.

1. $8 = 7\frac{\quad}{3}$
2. $4 = 3\frac{\quad}{6}$
3. $1 = \frac{\quad}{8}$
4. $6 = 5\frac{\quad}{4}$
5. $3 = 2\frac{\quad}{2}$
6. $9 = 8\frac{\quad}{10}$
7. $2 = 1\frac{\quad}{9}$
8. $5 = 4\frac{\quad}{7}$

Subtract.

9. $7 - 4\frac{3}{4}$
10. $10 - 5\frac{3}{6}$
11. $4 - 1\frac{2}{10}$
12. $6 - 2\frac{5}{8}$
13. $5 - 3\frac{1}{3}$
14. $4 - 3\frac{2}{10}$
15. $9 - 7\frac{4}{5}$
16. $2 - 1\frac{8}{12}$
17. $10 - 3\frac{2}{8}$
18. $7 - 1\frac{9}{12}$
19. $5 - 4\frac{3}{5}$
20. $6 - 5\frac{1}{2}$
21. $8 - 6\frac{6}{10}$
22. $12 - 10\frac{2}{6}$
23. $9 - 5\frac{3}{8}$

Find the difference.

24. $4 - \frac{1}{5} =$ ______
25. $1 - \frac{5}{6} =$ ______
26. $8 - 7\frac{2}{9} =$ ______
27. $6 - 2\frac{2}{3} =$ ______
28. $17 - 6\frac{7}{10} =$ ______
29. $5 - \frac{5}{6} =$ ______
30. $3 - \frac{6}{8} =$ ______
31. $10 - \frac{3}{9} =$ ______
32. $2 - \frac{8}{10} =$ ______

PROBLEM SOLVING

33. Charise had 10 ft of twine. She used $4\frac{3}{4}$ ft to tie up a package. Does she have enough twine left to tie up another package if she needs $6\frac{1}{4}$ ft of twine? ______________________

 Use with Lesson 5-10, text pages 182–183.

More Renaming in Subtraction

Name ______________________

Date ______________________

Subtract: $5\frac{1}{3} - 3\frac{5}{6} = \underline{\ ?\ }$

$5\frac{1}{3} = 5\frac{4}{12}$

$-3\frac{5}{6} = -3\frac{10}{12}$

$5\frac{4}{12} = 4 + 1 + \frac{4}{12}$

$= 4 + \frac{12}{12} + \frac{4}{12}$

$= 4 + \frac{16}{12}$

$4\frac{16}{12}$

$-3\frac{10}{12}$

$1\frac{6}{12} = 1\frac{1}{2}$

Complete.

1. $6\frac{3}{4} = 5 + 1 + \frac{3}{4}$
   $= 5 + \frac{4}{4} + \frac{3}{4}$
   $= 5\frac{\ \ }{4}$

2. $4\frac{7}{10} = \_\_ + 1 + \frac{7}{10}$
   $= \_\_ + \frac{10}{10} + \frac{7}{10}$
   $= \_\_\frac{\ \ }{10}$

3. $9\frac{5}{8} = 8 + \frac{\ \ }{8} + \frac{5}{8}$
   $= \_\_ + \frac{\ \ }{8} + \frac{5}{8}$
   $= \_\_ + \frac{\ \ }{8}$

Subtract.

4. $6\frac{3}{4} - 2\frac{7}{8}$

5. $5\frac{1}{6} - 2\frac{1}{2}$

6. $5\frac{1}{4} - \frac{3}{8}$

7. $9\frac{2}{3} - 7\frac{5}{6}$

8. $7\frac{1}{2} - \frac{4}{5}$

9. $4\frac{3}{4} - 3\frac{5}{6}$

10. $8\frac{1}{3} - 4\frac{7}{8}$

11. $3\frac{3}{9} - 1\frac{5}{6}$

12. $11\frac{1}{12} - 2\frac{5}{12}$

13. $8\frac{1}{4} - 1\frac{9}{14}$

14. $6\frac{5}{8} - 1\frac{4}{5}$

15. $7\frac{5}{9} - 6\frac{5}{6}$

16. $9\frac{5}{6} - 2\frac{7}{8}$

17. $12\frac{5}{7} - 9\frac{5}{6}$

18. $9\frac{1}{8} - 3\frac{5}{12}$

19. $8\frac{2}{3} - 6\frac{3}{4} =$ ______

20. $4\frac{1}{5} - \frac{4}{5} =$ ______

21. $3\frac{3}{8} - 2\frac{9}{10} =$ ______

22. $10\frac{1}{5} - \frac{12}{15} =$ ______

23. $7\frac{1}{2} - \frac{9}{12} =$ ______

24. $9\frac{1}{4} - \frac{5}{6} =$ ______

PROBLEM SOLVING

25. Mr. Lin drove $18\frac{1}{2}$ mi to work. Then he moved $6\frac{8}{10}$ mi closer to work. How far does he drive to work now? ______________________

Estimate to Compute

Name ______________________

Date ______________________

Round each fraction to the nearest whole number.
Then add or subtract.

Estimate the sum.

$8\frac{1}{3} + 2\frac{5}{7} + 11\frac{3}{10}$

$8 + 3 + 11 = 22$

Estimate the difference.

$9\frac{3}{4} - 2\frac{4}{5}$

$10 - 3 = 7$

Use rounding to estimate the sum.

1. $4\frac{7}{12} + 3\frac{2}{3}$
2. $8\frac{3}{10} + 2\frac{2}{5}$
3. $17\frac{5}{7} + 4\frac{1}{2}$
4. $13\frac{5}{8} + 4\frac{1}{8}$
5. $9\frac{2}{5} + 3\frac{4}{7}$
6. $16\frac{1}{2} + 4\frac{3}{4}$
7. $4\frac{1}{3} + 2\frac{7}{12}$
8. $8\frac{2}{5} + 4\frac{5}{6}$
9. $12\frac{1}{4} + 5\frac{3}{4}$
10. $6\frac{1}{8} + 2\frac{7}{11}$
11. $7\frac{2}{9} + 2\frac{1}{2}$
12. $4\frac{2}{3} + 9\frac{5}{8}$

Use rounding to estimate the difference.

13. $18\frac{7}{12} - 1\frac{5}{12}$
14. $9\frac{9}{10} - 4\frac{4}{5}$
15. $4\frac{4}{5} - \frac{2}{3}$
16. $13\frac{3}{4} - 6\frac{1}{2}$
17. $7\frac{1}{6} - 3\frac{5}{6}$
18. $6\frac{5}{9} - 2\frac{12}{19}$
19. $19\frac{4}{5} - 12\frac{7}{8}$
20. $16\frac{7}{9} - 4\frac{4}{11}$
21. $8\frac{4}{9} - 6\frac{2}{3}$
22. $11\frac{4}{5} - 2\frac{1}{5}$
23. $13\frac{7}{10} - 9\frac{4}{9}$
24. $17\frac{7}{9} - 6\frac{3}{10}$

Estimate. Use front-end estimation.

25. $12\frac{2}{5} + 3\frac{2}{3} + 3\frac{7}{9}$ ____
26. $14\frac{1}{2} + 5\frac{3}{11} + 2\frac{4}{13}$ ____
27. $9\frac{1}{7} + 8\frac{4}{5} + 3\frac{3}{8}$ ____
28. $18\frac{2}{3} + 7\frac{1}{6} + 3\frac{7}{10}$ ____
29. $22\frac{1}{3} - 17\frac{1}{2}$ ____
30. $14\frac{7}{11} - 4\frac{2}{11}$ ____
31. $8\frac{3}{5} - 1\frac{11}{12}$ ____
32. $19\frac{9}{10} - 4\frac{8}{9}$ ____
33. $17\frac{3}{8} - 8\frac{1}{7}$ ____

 Use with Lesson 5-12, text pages 186–187.

Problem-Solving Strategy: Working Backwards

Name ______________________

Date ______________________

Lisa bought a blouse for \$24.95 and a scarf for \$9.95.
She gave the clerk 2 bills and received \$5.10 change.
What bills did Lisa give the clerk?

Start with the change and work backwards.	\$5.10
Add the amount she paid for the scarf.	\$5.10 + \$9.95 = \$15.05
Add the amount she paid for the blouse.	\$15.05 + \$24.95 = \$40.00

Lisa gave \$40.00 to the clerk.
Two 20-dollar bills equal \$40.00.
Lisa gave two 20-dollar bills to the clerk.

Solve. Do your work on a separate sheet of paper.

1. Francine and Franklin baked some granola bars. The first customer bought $\frac{1}{2}$ dozen. The next customer bought 2 dozen. The third customer bought $1\frac{1}{2}$ dozen. There were 1 dozen left. How many granola bars did Francine and Franklin bake?

   ____

2. A bus leaves the school with some students on board. At the first stop, 4 students get off. At the second stop, 6 students get off. At each of the next 3 stops, 8 students get off. There are 10 students still on the bus. How many students were on the bus when it left the school?

   ____

3. Marlena says that if you add 8 to her age and then divide by 2, the answer is 10. How old is Marlena?

   ____

4. Sean says that if you double his age and then add 2, the answer is 30. How old is Sean?

   ____

5. Carmella spent \$6.95 for a book and \$1.50 for a birthday card. She had \$1.55 left. How much money did she have before she bought anything?

   ____

6. Brendan gathered seashells on the beach. He gave 7 to his younger sister and 7 to his older brother. He has 17 left. How many shells did he gather?

   ____

7. Gary baked some bran muffins. His mothers ate 1, his father ate 2, his sister and brother each ate 1, and his friends ate 8. There are 11 muffins left. How many did Gary bake?

   ____

8. Patty ties a bow on each basket she is making. She cuts 2 pieces of ribbon each $1\frac{1}{2}$ feet long and 2 pieces each $1\frac{3}{4}$ feet long. She has $8\frac{1}{2}$ feet of ribbon left. How much ribbon did she have in the beginning?

   ____

Multiplying Fractions

Name ____________________

Date ____________________

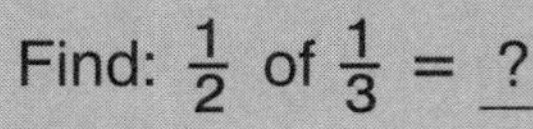

Find: $\frac{1}{2}$ of $\frac{1}{3}$ = ?

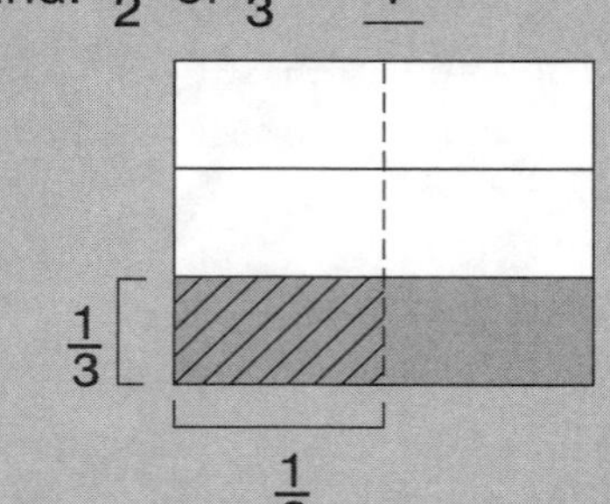

$\frac{1}{2}$ of $\frac{1}{3} = \frac{1}{6}$

Multiply: $\frac{2}{3} \times \frac{2}{5}$ = ?

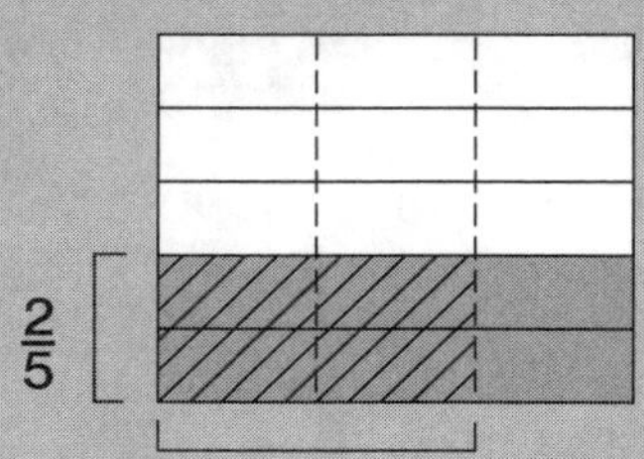

$\frac{2}{3} \times \frac{2}{5} = \frac{4}{15}$

Use the diagram to complete each statement.

1. 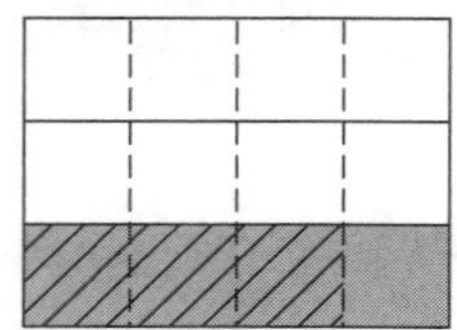

$\frac{3}{4}$ of $\frac{1}{3}$ = ____

2. 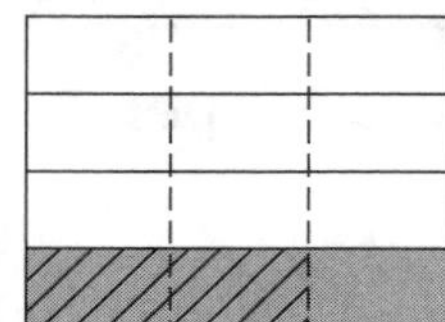

$\frac{2}{3}$ of $\frac{1}{4}$ = ____

3. 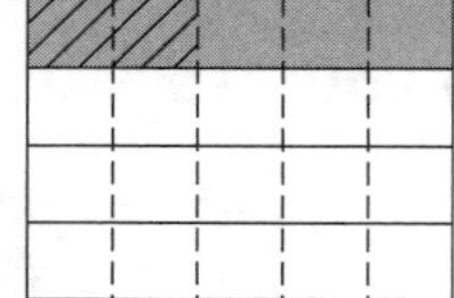

$\frac{2}{5}$ of $\frac{1}{4}$ = ____

4. 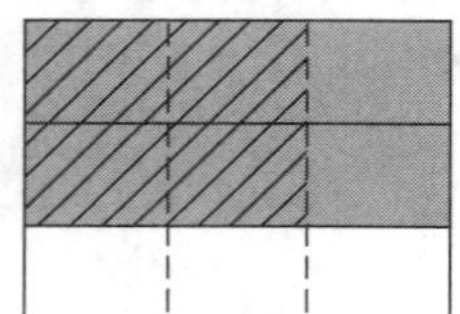

$\frac{2}{3} \times \frac{2}{3}$ = ____

5. 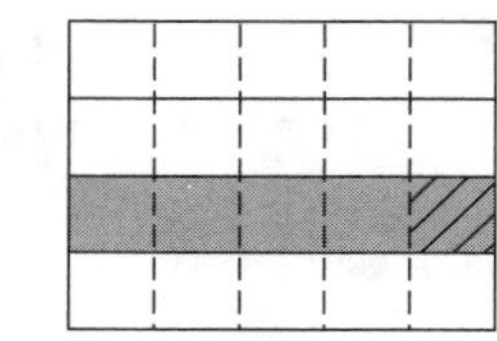

$\frac{1}{5} \times \frac{1}{4}$ = ____

6. 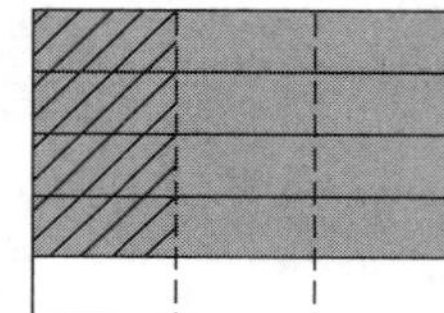

____ $\times \frac{4}{5} = \frac{4}{15}$

7. 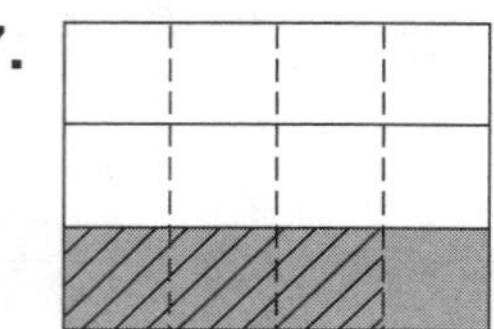

$\frac{3}{4} \times$ ____ $= \frac{3}{12}$

8. 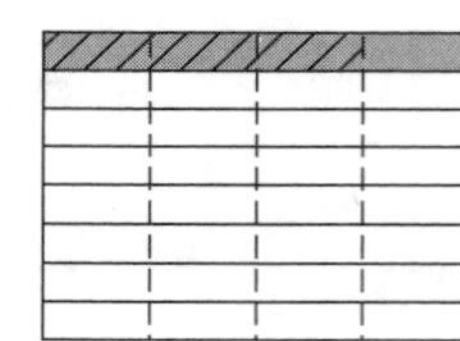

____ × ____ = ____

9. 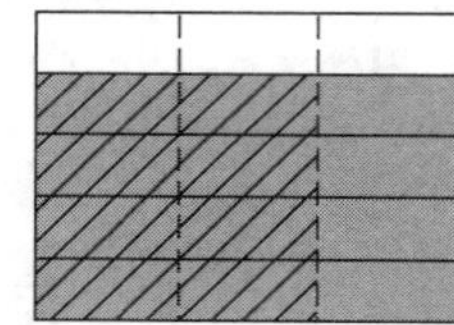

____ × ____ = ____

Find the diagram that matches each statement. Then complete.

10. $\frac{3}{4}$ of $\frac{2}{5}$ = ____

11. $\frac{1}{4}$ of $\frac{2}{3}$ = ____

12. $\frac{1}{3}$ of $\frac{1}{3}$ = ____

13. $\frac{2}{3}$ of $\frac{1}{2}$ = ____

a. 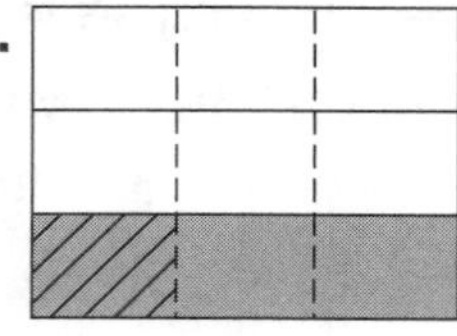

b. 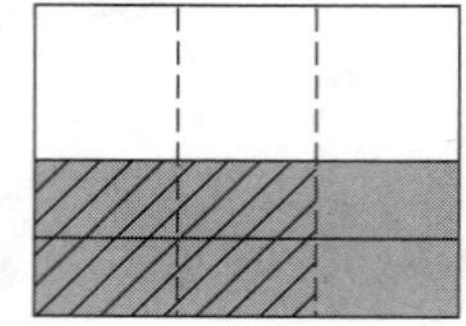

c. 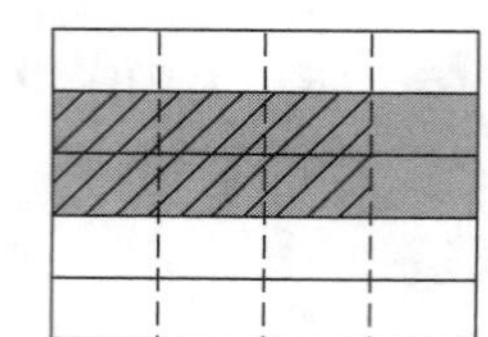

d. 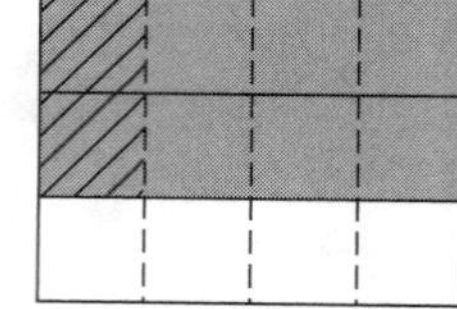

 Use with Lesson 6-1, text pages 198–199.

Multiplying Fractions by Fractions

Name ______________________

Date ______________________

Multiply: $\frac{2}{3} \times \frac{1}{4} = \underline{?}$

$$\frac{2}{3} \times \frac{1}{4} = \frac{2 \times 1}{3 \times 4} = \frac{2}{12}$$

$$= \frac{1}{6} \leftarrow \text{simplest form}$$

Multiply. Write each product in simplest form.

1. $\frac{1}{2} \times \frac{3}{4} =$ ____ **2.** $\frac{5}{6} \times \frac{1}{2} =$ ____ **3.** $\frac{1}{2} \times \frac{2}{5} =$ ____ **4.** $\frac{2}{3} \times \frac{1}{3} =$ ____

5. $\frac{3}{4} \times \frac{1}{3} =$ ____ **6.** $\frac{1}{3} \times \frac{1}{4} =$ ____ **7.** $\frac{4}{5} \times \frac{3}{4} =$ ____ **8.** $\frac{1}{4} \times \frac{2}{5} =$ ____

9. $\frac{3}{5} \times \frac{1}{2} =$ ____ **10.** $\frac{1}{9} \times \frac{2}{3} =$ ____ **11.** $\frac{2}{3} \times \frac{3}{4} =$ ____ **12.** $\frac{3}{7} \times \frac{1}{4} =$ ____

Compare. Write <, =, or >.

13. $\frac{1}{3} \times \frac{1}{5}$ ____ $\frac{1}{4} \times \frac{1}{3}$ **14.** $\frac{1}{2} \times \frac{1}{2}$ ____ $\frac{3}{4} \times \frac{1}{6}$ **15.** $\frac{1}{2} \times \frac{2}{3}$ ____ $\frac{2}{4} \times \frac{2}{3}$

16. $\frac{3}{4} \times \frac{1}{3}$ ____ $\frac{3}{5} \times \frac{1}{2}$ **17.** $\frac{3}{6} \times \frac{3}{6}$ ____ $\frac{1}{2} \times \frac{1}{2}$ **18.** $\frac{3}{4} \times \frac{3}{4}$ ____ $\frac{1}{2} \times \frac{5}{8}$

PROBLEM SOLVING

19. Jill took $\frac{2}{3}$ of the cookies she baked to school. She gave $\frac{1}{4}$ of these cookies to her teacher. What part of all the cookies did she give to her teacher? ______________

20. Conrad has a book about the outdoors, and $\frac{2}{5}$ of the book is about camping. Conrad has read $\frac{1}{3}$ of the the section on camping. What part of the book has Conrad read? ______________

21. Lucy created a design with different shapes. Stars made up $\frac{1}{4}$ of all the shapes in the design. Eight ninths of the stars are red. What fraction of all the shapes are red stars? ______________

22. Hector used $\frac{3}{4}$ of the money he earned to buy presents. He spent $\frac{1}{6}$ of the present money for his brother's gift. What part of his money did he spend on a gift for his brother? ______________

Multiplying Fractions and Whole Numbers

Name ____________________

Date ____________________

$5 \times \frac{2}{5} = \frac{5}{1} \times \frac{2}{5}$

$= \frac{5 \times 2}{1 \times 5}$

$= \frac{10}{5} = 2$

$3 \times \frac{2}{5} = \frac{3}{1} \times \frac{2}{5}$

$= \frac{3 \times 2}{1 \times 5}$

$= \frac{6}{5} = 1\frac{1}{5}$

$\frac{3}{4}$ of \$8 $= \frac{3}{4} \times \frac{8}{1}$

$= \frac{3 \times 8}{4 \times 1}$

$= \frac{24}{4} =$ \$6

Rename as an improper fraction with a denominator of 1.

1. 6 _____ **2.** 10 _____ **3.** 12 _____ **4.** 7 _____

Multiply.

5. $18 \times \frac{1}{3} =$ _____ **6.** $20 \times \frac{2}{5} =$ _____ **7.** $8 \times \frac{1}{4} =$ _____

8. $9 \times \frac{5}{9} =$ _____ **9.** $32 \times \frac{1}{4} =$ _____ **10.** $12 \times \frac{1}{3} =$ _____

11. $52 \times \frac{3}{4} =$ _____ **12.** $30 \times \frac{3}{10} =$ _____ **13.** $21 \times \frac{3}{7} =$ _____

14. $12 \times \frac{5}{6} =$ _____ **15.** $20 \times \frac{3}{5} =$ _____ **16.** $30 \times \frac{3}{5} =$ _____

17. $\frac{1}{5}$ of \$10 = _____ **18.** $\frac{3}{12}$ of \$36 = _____ **19.** $\frac{3}{4}$ of \$12 = _____

20. $\frac{2}{3}$ of \$18 = _____ **21.** $\frac{1}{3}$ of 5 = _____ **22.** $\frac{2}{5}$ of 8 = _____

23. $\frac{1}{2}$ of 19 = _____ **24.** $\frac{5}{6}$ of 9 = _____ **25.** $\frac{3}{7}$ of 63 = _____

PROBLEM SOLVING

26. Arthur walks $\frac{5}{8}$ mi to school. Jonathan rides a bus 8 times that far. How far does Jonathan ride to school? _____

27. A recipe to serve 4 people requires $\frac{2}{3}$ cup of flour. Jamika needs 5 times as much flour. How much flour does she need? _____

28. Delores read $\frac{2}{5}$ of the books on her reading list. There were 10 books on her list. How many books did Delores read? _____

29. Alex had 4 yd of cloth. He used $\frac{3}{8}$ of it to make a shirt. How many yards of cloth did Alex use to make the shirt? _____

 Use with Lesson 6-3, text pages 202–203.

Multiplying Using Cancellation

Name ____________________

Date ____________________

Multiply using cancellation.

$\frac{3}{4} \times \frac{8}{9} = \frac{3 \times \overset{2}{\cancel{8}}}{\underset{1}{\cancel{4}} \times 9}$ GCF of 4 and 8: 4 GCF of 3 and 9: 3

$= \frac{\overset{1}{\cancel{3}} \times \overset{2}{\cancel{8}}}{\underset{1}{\cancel{4}} \times \underset{3}{\cancel{9}}} = \frac{1 \times 2}{1 \times 3} = \frac{2}{3}$

$\frac{3}{4} \times 20 = \frac{3}{4} \times \frac{20}{1}$ GCF of 4 and 20: 4

$= \frac{3 \times \overset{5}{\cancel{20}}}{\underset{1}{\cancel{4}} \times 1} = \frac{3 \times 5}{1 \times 1} = \frac{15}{1} = 15$

Complete.

1. $\frac{4}{9} \times \frac{7}{24} = \frac{\overset{\square}{\cancel{4}} \times 7}{9 \times \underset{\square}{\cancel{24}}}$

$= \frac{1 \times}{9 \times} = \frac{}{}$

2. $\frac{1}{4} \times 12 = \frac{1}{\underset{\square}{\cancel{4}}} \times \frac{\overset{\square}{\cancel{12}}}{1}$

$= \frac{1 \times}{1 \times} = \frac{}{} =$ ________

Multiply using cancellation.

3. $\frac{2}{3} \times \frac{3}{5} =$ ______
4. $\frac{2}{5} \times \frac{1}{2} =$ ______
5. $\frac{7}{8} \times \frac{4}{7} =$ ______
6. $\frac{3}{5} \times \frac{5}{6} =$ ______
7. $\frac{2}{3} \times \frac{9}{10} =$ ______
8. $\frac{1}{2} \times \frac{4}{7} =$ ______
9. $\frac{6}{7} \times \frac{3}{10} =$ ______
10. $\frac{2}{5} \times \frac{5}{6} =$ ______
11. $\frac{1}{6} \times \frac{8}{9} =$ ______
12. $\frac{5}{12} \times \frac{6}{25} =$ ______
13. $\frac{4}{7} \times \frac{7}{10} =$ ______
14. $\frac{9}{10} \times \frac{10}{27} =$ ______
15. $\frac{5}{12} \times \frac{12}{35} =$ ______
16. $\frac{2}{3} \times 15 =$ ______
17. $\frac{3}{8} \times 32 =$ ______
18. $40 \times \frac{4}{5} =$ ______
19. $50 \times \frac{7}{10} =$ ______
20. $\frac{3}{4} \times 6 =$ ______
21. $8 \times \frac{5}{12} =$ ______
22. $\frac{5}{6} \times 9 =$ ______
23. $15 \times \frac{9}{10} =$ ______

PROBLEM SOLVING Use cancellation.

24. Donald bought 50 lb of grass seed. He used $\frac{4}{5}$ of it for a new lawn. How many pounds did he use? ____________

25. Ralph's dog weighs 24 lb and his cat weighs $\frac{5}{8}$ as much. How much does Ralph's cat weigh? ____________

Mixed Numbers to Improper Fractions

Name ______________________

Date ______________________

Rename $4\frac{1}{3}$ as an improper fraction.

$$4\frac{1}{3} = \frac{(3 \times 4) + 1}{3}$$
$$= \frac{12 + 1}{3}$$
$$= \frac{13}{3}$$

Think: $4\frac{1}{3}$ (+, ×)

Rename as an improper fraction.

1. $2\frac{1}{4} =$ ____	**2.** $6\frac{1}{2} =$ ____	**3.** $5\frac{1}{3} =$ ____	**4.** $3\frac{1}{8} =$ ____
5. $5\frac{3}{7} =$ ____	**6.** $3\frac{6}{7} =$ ____	**7.** $2\frac{7}{8} =$ ____	**8.** $7\frac{3}{5} =$ ____
9. $1\frac{5}{9} =$ ____	**10.** $6\frac{2}{3} =$ ____	**11.** $5\frac{3}{4} =$ ____	**12.** $4\frac{5}{6} =$ ____
13. $8\frac{3}{5} =$ ____	**14.** $13\frac{1}{3} =$ ____	**15.** $8\frac{1}{2} =$ ____	**16.** $15\frac{1}{2} =$ ____
17. $9\frac{5}{6} =$ ____	**18.** $12\frac{5}{7} =$ ____	**19.** $11\frac{3}{5} =$ ____	**20.** $10\frac{6}{7} =$ ____
21. $3\frac{1}{3} =$ ____	**22.** $4\frac{1}{2} =$ ____	**23.** $10\frac{2}{9} =$ ____	**24.** $7\frac{3}{4} =$ ____
25. $9\frac{2}{3} =$ ____	**26.** $1\frac{3}{8} =$ ____	**27.** $5\frac{3}{5} =$ ____	**28.** $6\frac{1}{7} =$ ____

PROBLEM SOLVING

29. A plumber needs to cut $7\frac{11}{16}$ inches from a piece of pipe. Write this length as an improper fraction. ____________

30. Daryl cuts a board that is $9\frac{1}{3}$ ft long. Write this length as an improper fraction. ____________

31. How would $3\frac{5}{6}$ be written as an improper fraction? ____________

32. A pencil $3\frac{5}{16}$ inches long. Write this length as an improper fraction. ____________

33. Lyall has a piece of rope that is $8\frac{4}{9}$ yd long. Write this length as an improper fraction. ____________

34. Jessica is $4\frac{3}{4}$ ft tall. Write her height as an improper fraction. ____________

35. Lamont jogs $2\frac{3}{10}$ mi everyday. Write this distance as an improper fraction. ____________

 Use with Lesson 6-5 text pages 206–207.

Multiplying Fractions and Mixed Numbers

Name ____________________

Date ____________________

Multiply: $\frac{3}{4} \times 1\frac{1}{5} = \underline{\ ?\ }$

$$\frac{3}{4} \times 1\frac{1}{5} = \frac{3}{4} \times \frac{6}{5}$$

$$= \frac{3 \times \overset{3}{\cancel{6}}}{\underset{2}{\cancel{4}} \times 5}$$

$$= \frac{3 \times 3}{2 \times 5} = \frac{9}{10}$$

Multiply: $7\frac{1}{2} \times \frac{4}{5} = \underline{\ ?\ }$

$$7\frac{1}{2} \times \frac{4}{5} = \frac{15}{2} \times \frac{4}{5}$$

$$= \frac{\overset{3}{\cancel{15}} \times \overset{2}{\cancel{4}}}{\underset{1}{\cancel{2}} \times \underset{1}{\cancel{5}}} = \frac{3 \times 2}{1 \times 1}$$

$$= \frac{6}{1} = 6$$

Write the product.

1. $2\frac{1}{4} \times \frac{1}{3} =$ ________
2. $3\frac{1}{3} \times \frac{1}{2} =$ ________
3. $\frac{1}{8} \times 2\frac{2}{5} =$ ________
4. $1\frac{1}{3} \times \frac{1}{6} =$ ________
5. $\frac{5}{6} \times 3\frac{1}{5} =$ ________
6. $7\frac{1}{2} \times \frac{1}{4} =$ ________
7. $\frac{1}{8} \times 3\frac{1}{4} =$ ________
8. $\frac{3}{8} \times 1\frac{5}{7} =$ ________
9. $\frac{3}{5} \times 3\frac{1}{2} =$ ________
10. $\frac{6}{7} \times 4\frac{2}{3} =$ ________
11. $\frac{3}{5} \times 1\frac{3}{5} =$ ________
12. $\frac{4}{9} \times 3\frac{3}{5} =$ ________

PROBLEM SOLVING

13. Frank lives $\frac{3}{10}$ mile from school. Michelle lives $2\frac{1}{3}$ times as far from school as Frank. How far does Michelle live from school? ________

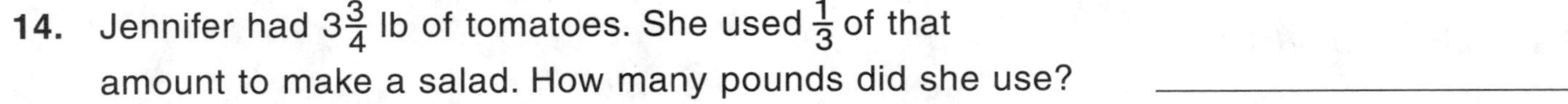

14. Jennifer had $3\frac{3}{4}$ lb of tomatoes. She used $\frac{1}{3}$ of that amount to make a salad. How many pounds did she use? ________

15. Daniel had $7\frac{3}{5}$ qt of strawberries. He used $\frac{1}{2}$ of that amount to make jam. How many quarts did he have left? ________

16. Shaya bought $2\frac{2}{3}$ lb of apples. She gave $\frac{1}{8}$ of that amount to her friend. How many pounds did she give to her friend? ________

17. Roger hiked $\frac{5}{7}$ mi to a lake. Then he hiked $3\frac{1}{2}$ times that distance to a mountain pass. How far is it from the lake to the mountain pass? ________

Multiplying Mixed Numbers

Name ______________________

Date ______________________

Multiply: $2\frac{1}{2} \times 2\frac{3}{5} = \underline{?}$

$$2\frac{1}{2} \times 2\frac{3}{5} = \frac{5}{2} \times \frac{13}{5}$$

$$= \frac{\overset{1}{\cancel{5}} \times 13}{2 \times \underset{1}{\cancel{5}}} = \frac{1 \times 13}{2 \times 1}$$

$$= \frac{13}{2} = 6\frac{1}{2}$$

Multiply: $6 \times 2\frac{1}{8} = \underline{?}$

$$6 \times 2\frac{1}{8} = \frac{6}{1} \times \frac{17}{8}$$

$$= \frac{\overset{3}{\cancel{6}} \times 17}{1 \times \underset{4}{\cancel{8}}} = \frac{3 \times 17}{1 \times 4}$$

$$= \frac{51}{4} = 12\frac{3}{4}$$

Multiply.

1. $1\frac{7}{8} \times 18 =$ ______________
2. $3\frac{3}{4} \times 4\frac{2}{3} =$ ______________
3. $7\frac{3}{10} \times 2\frac{1}{3} =$ ______________
4. $5\frac{1}{2} \times 2\frac{1}{2} =$ ______________
5. $3\frac{3}{8} \times 4\frac{1}{4} =$ ______________
6. $4\frac{2}{5} \times 25 =$ ______________
7. $2\frac{1}{3} \times 3 =$ ______________
8. $5\frac{1}{2} \times 5 =$ ______________
9. $4 \times 1\frac{1}{4} =$ ______________
10. $6 \times 1\frac{1}{5} =$ ______________
11. $2\frac{1}{3} \times 4\frac{2}{3} =$ ______________
12. $3\frac{1}{5} \times 1\frac{7}{8} =$ ______________
13. $1\frac{1}{3} \times 5\frac{1}{2} =$ ______________
14. $2\frac{2}{3} \times 3\frac{1}{4} =$ ______________

PROBLEM SOLVING

15. Midori jogs on weekdays. If she jogs $3\frac{1}{10}$ miles on each of the 5 days, what is her total distance per week? ______________

16. Emil jumped $4\frac{1}{2}$ ft in the high jump event at a track meet. Bill jumped $\frac{8}{9}$ as high as Emil. How high did Bill jump? ______________

17. Margaret used $3\frac{3}{4}$ yd of cloth to make curtains. She used $\frac{2}{5}$ of that amount to make a jacket. How much cloth did she use for the jacket? ______________

 Use with Lesson 6-7, text pages 210–211.

Division of Fractions

Name ______________________

Date ______________________

How many fourths in 2?

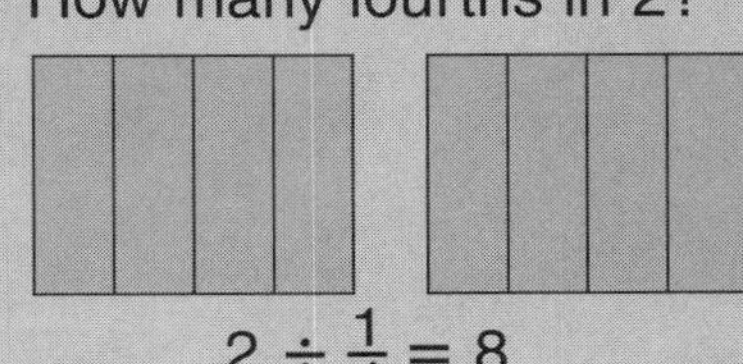

$2 \div \frac{1}{4} = 8$

How many twelfths in $\frac{5}{6}$?

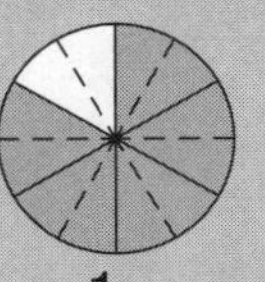

$\frac{5}{6} \div \frac{1}{12} = 10$

Use each diagram to find the quotient.

1.  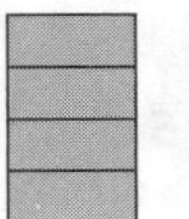 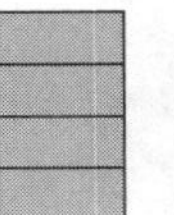 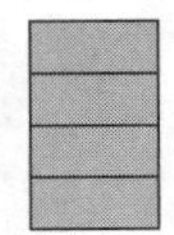

How many fourths in 4?
$4 \div \frac{1}{4} =$ ________

2. 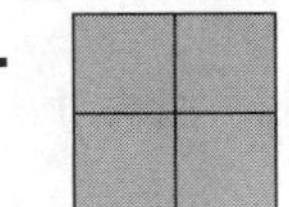 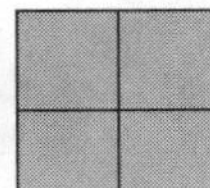 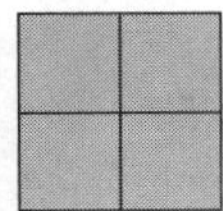

How many fourths in 3?
$3 \div \frac{1}{4} =$ ________

3. 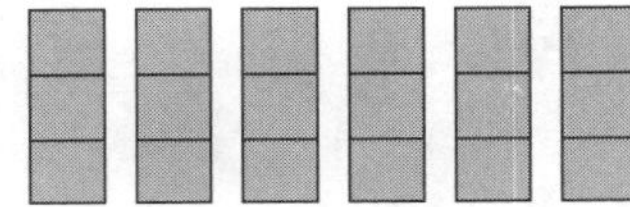

How many thirds in 6?
$6 \div \frac{1}{3} =$ ________

4. 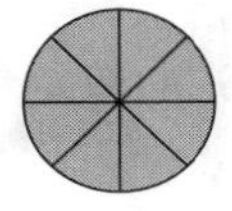 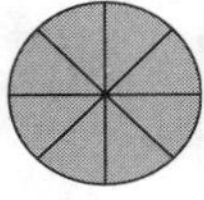 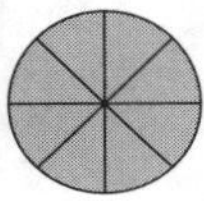

How many eighths in 3?
$3 \div \frac{1}{8} =$ ________

5. 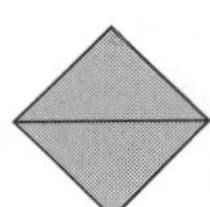

How many halves in 1?
$1 \div \frac{1}{2} =$ ________

6. 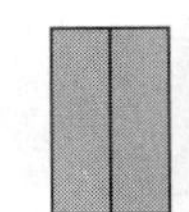   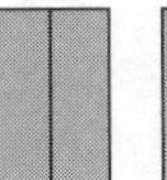

How many halves in 5?
$5 \div \frac{1}{2} =$ ________

7. 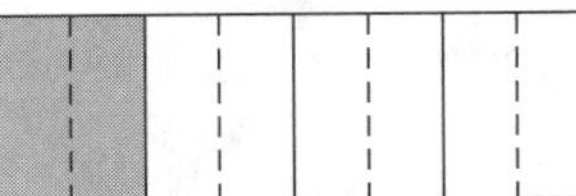

How many eighths in $\frac{1}{4}$?
$\frac{1}{4} \div \frac{1}{8} =$ ________

8. 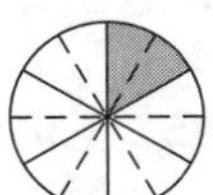

How many twelfths in $\frac{1}{6}$?
$\frac{1}{6} \div \frac{1}{12} =$ ________

9. 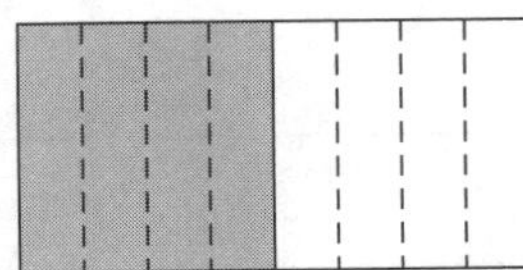

How many eighths in $\frac{1}{2}$?
$\frac{1}{2} \div \frac{1}{8} =$ ________

10. 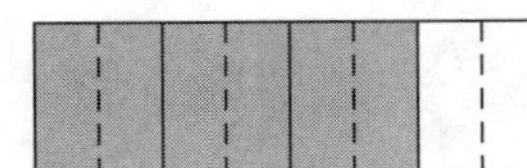

How many eighths in $\frac{3}{4}$?
$\frac{3}{4} \div \frac{1}{8} =$ ________

Reciprocals

Name ____________________

Date ____________________

Find the reciprocal.

$9 = \frac{9}{1}$ $\quad \frac{9}{1} \rightarrow \frac{1}{9}$

$\frac{\overset{1}{\cancel{9}}}{1} \times \frac{1}{\underset{1}{\cancel{9}}} = \frac{1}{1} = 1$

$\frac{1}{9}$ is the reciprocal of 9.

$\frac{2}{3} \rightarrow \frac{3}{2}$

$\frac{\overset{1}{\cancel{2}}}{\underset{1}{\cancel{3}}} \times \frac{\overset{1}{\cancel{3}}}{\underset{1}{\cancel{2}}} = \frac{1}{1} = 1$

$\frac{3}{2}$ is the reciprocal of $\frac{2}{3}$.

$1\frac{1}{4} = \frac{5}{4}$ $\quad \frac{5}{4} \rightarrow \frac{4}{5}$

$\frac{\overset{1}{\cancel{5}}}{\underset{1}{\cancel{4}}} \times \frac{\overset{1}{\cancel{4}}}{\underset{1}{\cancel{5}}} = \frac{1}{1} = 1$

$\frac{4}{5}$ is the reciprocal of $1\frac{1}{4}$.

Write the missing reciprocal in each statement.

1. $8 \times$ _____ $= 1$ **2.** $3 \times$ _____ $= 1$ **3.** $\frac{1}{2} \times$ _____ $= 1$ **4.** $\frac{1}{5} \times$ _____ $= 1$

5. $\frac{7}{8} \times$ _____ $= 1$ **6.** $\frac{4}{5} \times$ _____ $= 1$ **7.** $1\frac{1}{2} \times$ _____ $= 1$ **8.** $2\frac{1}{4} \times$ _____ $= 1$

Are the numbers reciprocals? Write *Yes* or *No*.

9. $12, \frac{1}{12}$ _____ **10.** $\frac{1}{3}, \frac{2}{6}$ _____ **11.** $\frac{4}{9}, \frac{9}{4}$ _____ **12.** $\frac{5}{8}, \frac{8}{5}$ _____

13. $\frac{1}{10}, 10$ _____ **14.** $2\frac{1}{5}, \frac{5}{11}$ _____ **15.** $3\frac{1}{2}, \frac{2}{5}$ _____ **16.** $1\frac{7}{8}, \frac{8}{15}$ _____

Write the reciprocal of each number.

17. 1 _____ **18.** 20 _____ **19.** $\frac{1}{3}$ _____ **20.** $\frac{1}{8}$ _____

21. $\frac{9}{10}$ _____ **22.** $\frac{7}{12}$ _____ **23.** $\frac{10}{3}$ _____ **24.** $\frac{9}{5}$ _____

25. $\frac{14}{9}$ _____ **26.** $1\frac{1}{6}$ _____ **27.** $2\frac{3}{4}$ _____ **28.** $3\frac{1}{5}$ _____

PROBLEM SOLVING Use the numbers in the box.

$\frac{1}{3}$	$\frac{5}{2}$	$\frac{9}{4}$	$\frac{7}{10}$

29. Write the fractions that are less than 1. Then write their reciprocals. ____________________

30. Write the fractions that are greater than 1. Then write their reciprocals. ____________________

31. What number times $\frac{5}{16}$ equals 1? ____________________

32. What number times 100 equals 1? ____________________

33. Use the numbers 7 and 11 to write a multiplication sentence with a product of 1. ____________________

Use with Lesson 6-9, text pages 214–215.

Dividing Whole Numbers by Fractions

Name ____________________

Date ____________________

Divide: $6 \div \frac{3}{8} = \underline{?}$

$$6 \div \frac{3}{8} = \frac{6}{1} \div \frac{3}{8}$$

$$= \frac{6}{1} \times \frac{8}{3}$$

$$= \frac{\overset{2}{\cancel{6}} \times 8}{1 \times \underset{1}{\cancel{3}}} = \frac{16}{1} = 16$$

Multiply by the reciprocal of the divisor.

Divide: $7 \div \frac{2}{5} = \underline{?}$

$$7 \div \frac{2}{5} = \frac{7}{1} \div \frac{2}{5}$$

$$= \frac{7}{1} \times \frac{5}{2}$$

$$= \frac{7 \times 5}{1 \times 2} = \frac{35}{2} = 17\frac{1}{2}$$

Complete.

1. $6 \div \frac{1}{4} = \frac{6}{1} \div \frac{1}{4}$

   $= \frac{6}{1} \times$ ___

   $=$ ____________________

2. $3 \div \frac{2}{7} = \frac{3}{1} \div \frac{2}{7}$

   $= \frac{3}{1} \times$ ___

   $=$ ____________________

Divide.

3. $12 \div \frac{2}{3} =$ __________
4. $16 \div \frac{5}{8} =$ __________
5. $6 \div \frac{1}{3} =$ __________
6. $10 \div \frac{1}{3} =$ __________
7. $9 \div \frac{4}{5} =$ __________
8. $18 \div \frac{3}{4} =$ __________
9. $8 \div \frac{2}{5} =$ __________
10. $21 \div \frac{2}{5} =$ __________
11. $1 \div \frac{1}{3} =$ __________
12. $25 \div \frac{1}{8} =$ __________
13. $2 \div \frac{7}{10} =$ __________
14. $15 \div \frac{3}{4} =$ __________
15. $7 \div \frac{2}{5} =$ __________
16. $14 \div \frac{1}{3} =$ __________

PROBLEM SOLVING

17. Lee needs pieces of wire that are each $\frac{2}{5}$ ft long. How many pieces can he cut from a 6-ft length of wire? ____________________

18. A pie is divided into eight equal pieces. How many pieces would there be if it were divided into pieces only $\frac{1}{2}$ that size? ____________________

Dividing Fractions by Fractions

Name ______________________

Date ______________________

Divide: $\frac{3}{4} \div \frac{1}{8} = \underline{?}$

$$\frac{3}{4} \div \frac{1}{8} = \frac{3}{4} \times \frac{8}{1}$$

$$= \frac{3 \times \overset{2}{\cancel{8}}}{\underset{1}{\cancel{4}} \times 1} = \frac{3 \times 2}{1 \times 1}$$

$$= \frac{6}{1} = 6$$

Divide: $\frac{4}{5} \div \frac{2}{3} = \underline{?}$

$$\frac{4}{5} \div \frac{2}{3} = \frac{4}{5} \times \frac{3}{2}$$

$$= \frac{\overset{2}{\cancel{4}} \times 3}{5 \times \underset{1}{\cancel{2}}} = \frac{2 \times 3}{5 \times 1}$$

$$= \frac{6}{5} = 1\frac{1}{5}$$

Complete.

1. $\frac{2}{3} \div \frac{2}{9} = \frac{2}{3} \times \frac{9}{2} =$ ____________
2. $\frac{4}{5} \div \frac{1}{10} = \frac{4}{5} \times \frac{10}{1} =$ ____________
3. $\frac{7}{8} \div \frac{3}{4} = \frac{7}{8} \times \frac{4}{3} =$ ____________
4. $\frac{5}{8} \div \frac{1}{3} = \frac{5}{8} \times \frac{3}{1} =$ ____________
5. $\frac{2}{7} \div \frac{4}{7} = \frac{2}{7} \times \frac{7}{4} =$ ____________
6. $\frac{3}{5} \div \frac{9}{10} = \frac{3}{5} \times \frac{10}{9} =$ ____________

Divide.

7. $\frac{1}{2} \div \frac{1}{10} =$ ____________
8. $\frac{1}{4} \div \frac{1}{8} =$ ____________
9. $\frac{5}{6} \div \frac{2}{3} =$ ____________
10. $\frac{7}{8} \div \frac{1}{2} =$ ____________
11. $\frac{2}{5} \div \frac{4}{15} =$ ____________
12. $\frac{7}{12} \div \frac{1}{3} =$ ____________
13. $\frac{1}{9} \div \frac{2}{3} =$ ____________
14. $\frac{3}{8} \div \frac{1}{2} =$ ____________
15. $\frac{2}{7} \div \frac{6}{7} =$ ____________
16. $\frac{4}{5} \div \frac{9}{10} =$ ____________

PROBLEM SOLVING

17. How many glasses that each contain $\frac{1}{8}$ liter of milk can be poured from a half liter of milk? ____________

18. Tom has a board $\frac{2}{3}$ yd long. How many $\frac{1}{8}$-yd-long sections can he cut from the board? ____________

 Use with Lesson 6-11, text pages 218–219.

Dividing Fractions by Whole Numbers

Name ______________________

Date ______________________

Divide: $\frac{1}{5} \div 2 = \underline{?}$

$\frac{1}{5} \div 2 = \frac{1}{5} \div \frac{2}{1}$

$= \frac{1}{5} \times \frac{1}{2}$

$= \frac{1 \times 1}{5 \times 2} = \frac{1}{10}$

Divide: $\frac{2}{7} \div 4 = \underline{?}$

$\frac{2}{7} \div 4 = \frac{2}{7} \div \frac{4}{1}$

$= \frac{2}{7} \times \frac{1}{4} = \frac{\overset{1}{\cancel{2}} \times 1}{7 \times \underset{2}{\cancel{4}}}$

$= \frac{1 \times 1}{7 \times 2} = \frac{1}{14}$

Complete.

1. $\frac{1}{2} \div 3 = \frac{1}{2} \div \frac{3}{1}$
   $= \frac{1}{2} \times \frac{1}{3} =$ ______________

2. $\frac{3}{5} \div 15 = \frac{3}{5} \div \frac{15}{1}$
   $= \frac{3}{5} \times \frac{1}{15} =$ ______________

Divide.

3. $\frac{1}{2} \div 5 =$ ______________
4. $\frac{2}{3} \div 5 =$ ______________
5. $\frac{5}{8} \div 3 =$ ______________
6. $\frac{4}{7} \div 2 =$ ______________
7. $\frac{2}{3} \div 6 =$ ______________
8. $\frac{1}{4} \div 4 =$ ______________
9. $\frac{3}{8} \div 3 =$ ______________
10. $\frac{7}{10} \div 2 =$ ______________
11. $\frac{1}{2} \div 7 =$ ______________
12. $\frac{9}{10} \div 5 =$ ______________
13. $\frac{4}{5} \div 8 =$ ______________
14. $\frac{7}{16} \div 14 =$ ______________

PROBLEM SOLVING

15. Mrs. Jamison has a garden that is $\frac{1}{3}$ of an acre. She wants to divide it into 6 equal sections. What part of an acre will each section be? ______________

16. During the fair, the fifth grade class uses $\frac{1}{4}$ of the gym. They divide the space into 5 equal sections for displays. What part of the gym is used for each display? ______________

Dividing Mixed Numbers by Fractions

Name ______________________

Date ______________________

Divide: $3\frac{1}{5} \div \frac{1}{10} = \underline{?}$

$$3\frac{1}{5} \div \frac{1}{10} = \frac{16}{5} \div \frac{1}{10} = \frac{16}{5} \times \frac{10}{1}$$

$$= \frac{16 \times \overset{2}{\cancel{10}}}{\underset{1}{\cancel{5}} \times 1} = \frac{16 \times 2}{1 \times 1}$$

$$= \frac{32}{1} = 32$$

Divide: $2\frac{1}{3} \div \frac{2}{5} = \underline{?}$

$$2\frac{1}{3} \div \frac{2}{5} = \frac{7}{3} \div \frac{2}{5}$$

$$= \frac{7}{3} \times \frac{5}{2}$$

$$= \frac{35}{6} = 5\frac{5}{6}$$

Divide.

1. $2\frac{1}{2} \div \frac{1}{4} =$ ______________
2. $1\frac{5}{6} \div \frac{1}{12} =$ ______________
3. $6\frac{1}{2} \div \frac{1}{2} =$ ______________
4. $7\frac{3}{4} \div \frac{1}{4} =$ ______________
5. $1\frac{1}{2} \div \frac{1}{3} =$ ______________
6. $3\frac{1}{3} \div \frac{1}{4} =$ ______________
7. $2\frac{1}{2} \div \frac{4}{9} =$ ______________
8. $6\frac{1}{3} \div \frac{2}{5} =$ ______________
9. $2\frac{1}{3} \div \frac{4}{9} =$ ______________
10. $7\frac{1}{5} \div \frac{8}{15} =$ ______________

PROBLEM SOLVING

11. Chris needs pieces of yarn $\frac{1}{2}$ dm long to use as strands of hair for a rag doll she is making. She has a $16\frac{1}{2}$-dm length of yarn. How many strands of hair can she make? ______________

12. A shop teacher had a board that was $8\frac{1}{2}$ feet long. He cut it into $\frac{1}{2}$-ft sections. How many sections were there? ______________

13. Jenna is making a scarf that is $3\frac{3}{4}$ feet long. If there is a tassel every $\frac{3}{4}$ ft, how many tassels are on one long side of the scarf? ______________

14. Malik notices trail markers every $\frac{3}{10}$ mi. If he hikes $4\frac{1}{2}$ mi, how many trail markers will he see? ______________

Dividing Mixed Numbers

Name ____________________

Date ____________________

Divide: $2\frac{1}{2} \div 3\frac{1}{3} = \underline{?}$

$$2\frac{1}{2} \div 3\frac{1}{3} = \frac{5}{2} \div \frac{10}{3}$$

$$= \frac{5}{2} \times \frac{3}{10}$$

$$= \frac{\overset{1}{\cancel{5}} \times 3}{2 \times \underset{2}{\cancel{10}}} = \frac{1 \times 3}{2 \times 2}$$

$$= \frac{3}{4}$$

Divide: $7 \div 1\frac{3}{4} = \underline{?}$

$$7 \div 1\frac{3}{4} = \frac{7}{1} \div \frac{7}{4}$$

$$= \frac{7}{1} \times \frac{4}{7}$$

$$= \frac{\overset{1}{\cancel{7}} \times 4}{1 \times \underset{1}{\cancel{7}}} = \frac{1 \times 4}{1 \times 1}$$

$$= 4$$

Divide.

1. $4\frac{1}{4} \div 2\frac{1}{8} =$ ____________
2. $1\frac{1}{2} \div 4\frac{1}{2} =$ ____________
3. $5 \div 1\frac{2}{3} =$ ____________
4. $7 \div 1\frac{2}{5} =$ ____________
5. $3\frac{1}{2} \div \frac{1}{3} =$ ____________
6. $5\frac{1}{8} \div \frac{1}{16} =$ ____________
7. $5\frac{1}{2} \div 1\frac{2}{3} =$ ____________
8. $8\frac{1}{8} \div \frac{1}{8} =$ ____________
9. $4\frac{2}{5} \div 2\frac{3}{4} =$ ____________
10. $5 \div 6\frac{1}{4} =$ ____________
11. $3\frac{3}{4} \div 5 =$ ____________
12. $12 \div 1\frac{1}{3} =$ ____________
13. $4\frac{3}{4} \div 2\frac{1}{4} =$ ____________
14. $2\frac{1}{5} \div 1\frac{1}{10} =$ ____________

PROBLEM SOLVING

15. Charlie is cutting wire into $2\frac{1}{4}$-ft pieces. How many pieces can he cut from a 45-ft roll? ____________

16. Pablo took $9\frac{1}{2}$ hours to paint 5 pictures. How long did it take him to paint each picture if he spent the same amount of time on each? ____________

17. Loren studied for $3\frac{1}{3}$ h. She took a break every $\frac{2}{3}$ h. How many breaks did she take? ____________

Estimating with Mixed Numbers

Name ______________________

Date ______________________

Estimate each product.

$1\frac{2}{3} \times 4\frac{1}{6}$ → $2 \times 4 = 8$

$3\frac{1}{8} \times 5$ → $3 \times 5 = 15$

$6 \times 2\frac{3}{4}$ → $6 \times 3 = 18$

Estimate each product.

1. $7 \times 2\frac{1}{4}$ ________
2. $5 \times 3\frac{1}{2}$ ________
3. $6 \times 1\frac{2}{3}$ ________
4. $5 \times 2\frac{3}{8}$ ________
5. $4\frac{1}{5} \times 1\frac{3}{7}$ ________
6. $2\frac{5}{8} \times 6\frac{9}{10}$ ________
7. $3\frac{2}{9} \times 7\frac{5}{6}$ ________
8. $8\frac{1}{8} \times 7\frac{5}{6}$ ________
9. $7\frac{1}{3} \times 2$ ________
10. $3\frac{3}{4} \times 7$ ________
11. $2\frac{1}{6} \times 8$ ________
12. $6\frac{1}{2} \times 9$ ________
13. $3\frac{4}{5} \times 2\frac{2}{3}$ ________
14. $4\frac{5}{7} \times 6\frac{1}{8}$ ________
15. $8\frac{1}{4} \times 5\frac{1}{2}$ ________
16. $3\frac{4}{9} \times 7\frac{1}{5}$ ________
17. $9\frac{1}{2} \times 4\frac{1}{3}$ ________
18. $2\frac{2}{5} \times 7\frac{3}{4}$ ________
19. $4 \times 8\frac{3}{10}$ ________
20. $5\frac{8}{9} \times 5$ ________
21. $14\frac{2}{5} \times 2\frac{3}{10}$ ________
22. $10 \times 2\frac{2}{3}$ ________
23. $4\frac{7}{8} \times 9$ ________
24. $6\frac{3}{4} \times 5\frac{1}{6}$ ________

Use compatible numbers to estimate. Then write whether the actual product is *less than* or *greater than* the estimated product.

25. $\frac{1}{2} \times 19\frac{1}{4}$

____

26. $\frac{3}{4} \times 24\frac{3}{8}$

____

27. $2\frac{3}{5} \times 18$

____

28. $1\frac{1}{7} \times 29$

____

Use with Lesson 6-15, text pages 226–227.

Problem-Solving Strategy: Use Simpler Numbers

Name ______________________

Date ______________________

Harvey lives $\frac{3}{4}$ mile from school. He walks to school each school-day morning and back home again in the afternoon. How many miles does Harvey walk in $2\frac{1}{2}$ days?

NUMBER OF MILES A DAY		NUMBER OF DAYS		TOTAL MILES
$2 \times \frac{3}{4}$	$\times$	$2\frac{1}{2}$	$=$	?
$\frac{2}{1} \times \frac{3}{4}$	$\times$	$\frac{5}{2}$	$=$	?
$\frac{\not{2}^{1}}{1} \times \frac{3}{\not{4}_{2}}$	$\times$	$\frac{5}{2}$	$=$	$\frac{15}{4}$
			$=$	$3\frac{3}{4}$

Use simpler numbers. If Harvey lived 1 mile from school and you want to find how many miles he walks in 3 days, multiply $3 \times (2 \times 1)$ to find the answer.
So find the number of miles he walks each day and then multiply by the number of days.

Harvey walks $3\frac{3}{4}$ miles in $2\frac{1}{2}$ days.

Solve. Do your work on a separate sheet of paper.

1. How many hours will it take Wendy to make $4\frac{2}{3}$ batches of cookies if she can make $1\frac{1}{3}$ batches in 1 hour?

____

2. Terrell drove at a speed of $48\frac{1}{2}$ miles per hour for $2\frac{1}{4}$ hours. How far did he travel?

____

3. Ronny unpacks pears and stacks them in a bin at the grocery store. Each carton contains $3\frac{1}{2}$ dozen pears. How many cartons will Ronny unpack if he fills a bin that holds $10\frac{1}{2}$ dozen pears?

____

4. Nina practices her guitar every week-night for $\frac{3}{4}$ hour. On Saturday and Sunday, she practices $1\frac{1}{4}$ hours each day. How many hours does Nina practice in 2 weeks?

____

5. It takes Leon $\frac{1}{3}$ h to ride his bicycle from his home to John's home. One day, after he arrived at John's home, Leon realized he had forgotten to bring a book. So he returned home to get the book and then rode back to John's. After visiting, he rode home again. How many hours did Leon ride?

____

6. A recipe for fruit punch calls for $2\frac{1}{2}$ c of orange juice, $1\frac{1}{3}$ c of pineapple juice, and $1\frac{3}{4}$ c of soda water. If June makes 4 times the recipe, how many cups of fruit punch will she have?

____

Probability

Name ____________________

Date ____________________

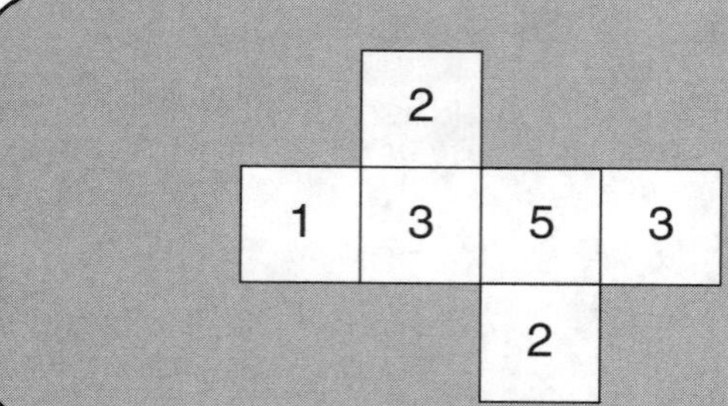

$P(1) = \frac{1}{6}$ ← 1

$P(2) = \frac{2}{6} = \frac{1}{3}$ ← 2, 2

$P(2 \text{ or } 3) = \frac{4}{6} = \frac{2}{3}$ ← 2, 2, 3, 3

$P(\text{not } 3) = \frac{4}{6} = \frac{2}{3}$ ← 1, 2, 2, 5

Use the number cube. Find the probability of each event.

1a. $P(2)$ = ________

b. $P(4)$ = ________

c. $P(3)$ = ________

d. $P(\textit{not } 3)$ = ________

e. $P(\textit{not } 4)$ = ________

f. $P(\textit{not } 2)$ = ________

2. In 600 rolls, predict how many times you will roll

a. 2 ________ **b.** 3 ________ **c.** 4 ________

Use the spinner at the right to find the probability of each event.

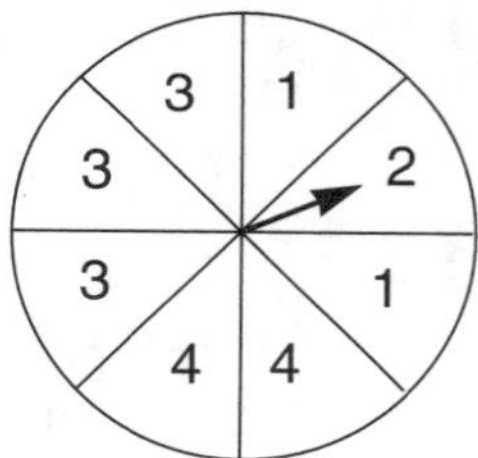

3. $P(3)$ = ________

4. $P(8)$ = ________

5. $P(4)$ = ________

6. $P(2)$ = ________

7. $P(1)$ = ________

8. $P(\textit{not } 1)$ = ________

A bag contains 2 black marbles, 3 white marbles, 4 green marbles, 5 red marbles, and 6 blue marbles. Find each probability.

9. $P(\textit{not}$ black$)$ = ________

10. $P($orange$)$ = ________

11. $P($red or blue$)$ = ________

12. $P($red or green$)$ = ________

13. $P($black or white$)$ = ________

14. $P(\textit{not}$ red$)$ = ________

15. $P($white or red$)$ = ________

16. $P(\textit{not}$ blue$)$ = ________

17. $P($black or red$)$ = ________

18. $P(\textit{not}$ purple$)$ = ________

PROBLEM SOLVING

18. Tim has 2 dimes, 1 nickel, and 1 quarter in his pocket. He picks out a coin at random. What is the probability that the coin is worth exactly 10¢?

 Use with Lesson 7-1, text pages 238–239.

Tree Diagrams

Name ______________________

Date ______________________

Toss a coin and spin the spinner. Find: $P(H, 3) =$?

$P(H, 3) = \frac{1}{6}$

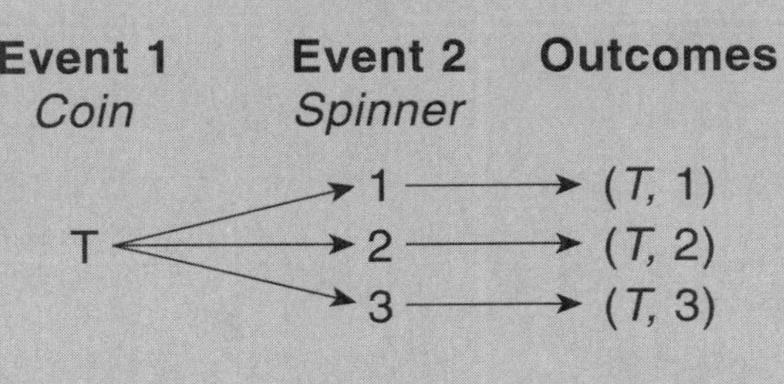

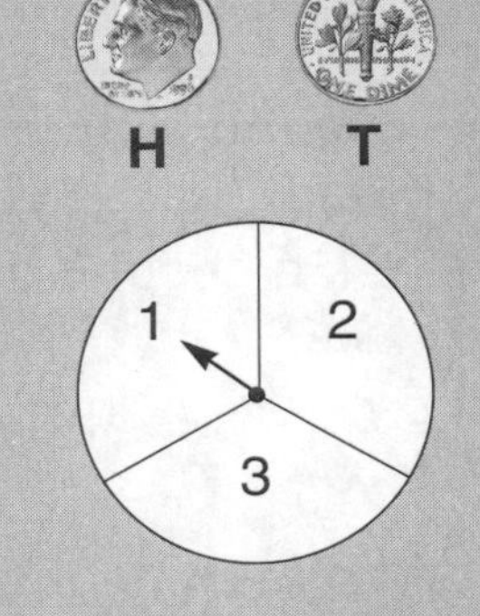

A

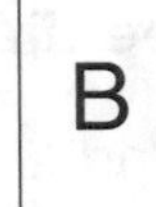

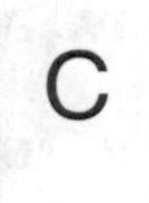

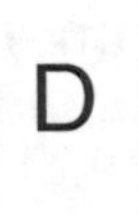

Find each probability. Use the spinner and the cards above.

1. $P(2, A)$ ________ **2.** $P(\text{odd}, B)$ ________ **3.** $P(\text{even}, D)$ ________ **4.** $P(1, A \text{ or } C)$ ________ **5.** $P(\textit{not}\ 2, \textit{not}\ A)$ ________

Draw a tree diagram on a separate sheet of paper.
Then find the probability.

6. $P(H, 5)$ ________ **7.** $P(T, 2)$ ________

8. $P(T, \text{even})$ ________ **9.** $P(H, \text{odd})$ ________

10. $P(T, 2 \text{ or } 3)$ ________ **11.** $P(H, \textit{not}\ 2)$ ________

Heads (*H*)

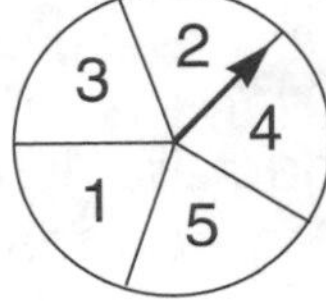

PROBLEM SOLVING You may draw a tree diagram on a separate sheet of paper.

12. Candace has 5 T-shirts: 1 pink, 2 blue, 1 red, and 1 white. She also has 2 hats: 1 white and 1 blue. If she selects a shirt and a hat without looking, what is the probability that she will get a pink shirt and a white hat? ______________________

13. Derek has 2 pairs of socks in his drawer: 1 blue pair and one brown pair. He also has 2 pairs of shoes: 1 black pair and 1 brown pair. If he picks 1 pair of socks and 1 pair of shoes at random, what is he probability that he picks socks and shoes of the same color? ______________________

Independent and Dependent Events

Name ______________________

Date ______________________

A bank contains 1 penny (p), 1 nickel (n), and 1 dime (d).

Independent Events:
Pick a coin. Return it.
Pick a second coin.

$p \to p, n, d \quad n \to p, n, d \quad d \to p, n, d$

$P(p, n) = \frac{1}{9}$

Dependent Events:
Pick a coin. *Do not* return it.
Pick a second coin.

$p \to n, d \quad n \to p, d \quad d \to p, n$

$P(p, n) = \frac{1}{6}$

Find the probability. You may draw a tree diagram.

A bank contains 1 nickel (n), 1 dime (d) and 2 quarters (q).

1. Pick a coin. Put it back. Then pick another coin.

a. $P(n, d) =$ ______ **b.** $P(q, d) =$ ______ **c.** $P(q, q) =$ ______

d. $P(\text{not } n, d) =$ ______ **e.** $P(q, \text{not } d) =$ ______ **f.** $P(n, \text{not } q) =$ ______

2. Pick a coin. Do *not* put it back. Then pick another coin.

a. $P(n, d) =$ ______ **b.** $P(q, d) =$ ______ **c.** $P(q, q) =$ ______

d. $P(\text{not } n, d) =$ ______ **e.** $P(q, \text{not } d) =$ ______ **f.** $P(n, \text{not } q) =$ ______

In a bag are 1 red marble (R), 1 blue marble (B), 1 white marble (W), and 1 green marble (G).

3. Pick a marble. Put it back. Then pick another marble.

a. $P(W, G) =$ ______ **b.** $P(R, \text{not } B) =$ ______ **c.** $P(\text{not } G, R) =$ ______

4. Pick a marble. Do *not* put it back. Then pick another marble.

a. $P(W, G) =$ ______ **b.** $P(R, \text{not } B) =$ ______ **c.** $P(\text{not } G, R) =$ ______

PROBLEM SOLVING

5. Suppose there are 2 dimes and 3 quarters in a bank. You shake it so that 2 coins fall out. What is the probability that

a. both coins are quarters? ______ **b.** both coins are dimes? ______

c. both coins are the same? ______ **d.** you shake out one of each coin? ______

Use with Lesson 7-3, text pages 242–243.

Finding Averages

Name ____________________

Date ____________________

Find the average, or mean, of 73, 64, 76, and 68.

Add.

$73 + 64 + 76 + 68 = 281$

The average of 73, 64, 76, and 68 is $70\frac{1}{4}$.

Divide.

$4\overline{)281}$ = 70 R1 $= 70\frac{1}{4}$

Check.

$70\frac{1}{4} \times 4 = \frac{281}{4} \times 4$

$= 281$

Find the average and check.

1. 72, 73, 74 ________

2. 51, 50, 45, 42 ________

3. 12, 9, 15, 21, 18 ________

4. 38, 32, 33, 34, 36 ________

Find the average for each set of data. Do your work on a separate sheet of paper.

5. Ages of brothers and sisters: 7, 8, 9, 10, 11, 12, 13 ________

6. Costs of motel rooms for one night: \$35, \$27, \$40, \$38, \$48, \$46 ________

7. Heights of trees: 26 ft, 25 ft, 30 ft, 27 ft, 28 ft, 32 ft ________

8.

Months	Sales Totals
March–April	\$16,555
May–June	\$20,138
July–August	\$25,892
Sept.–October	\$14,023

____

9.

Mountain	Height
Liberty Cap	14,112 ft
Pikes Peak	14,110 ft
Snowmass	14,092 ft
Vancouver	15,700 ft

____

PROBLEM SOLVING

10. Lambros's scores on his math tests were 78, 91, 85, 84, 97, and 87. What is his average grade? ________

11. The same mug was sold at different stores for \$4.00, \$5.00, \$4.50, \$4.75, and \$3.50. What was the average price of the mug? ________

12. Alison played a video game. Her scores were 121, 143, 90, and 162. What was her average score? ________

13. Ms. Peterson sold 4 automobiles. The total amount of the sales was \$70,736. What was the average price of the cars? ________

Collecting, Organizing, and Working with Data

Name ______________________

Date ______________________

Favorite Pet		
Pet	**Tally**	**Total**
Horse	~~IIII~~ II	7
Cat	~~IIII~~ ~~IIII~~	10
Dog	~~IIII~~ ~~IIII~~	10
Bird	~~IIII~~	5

Range: $10 - 5 = 5$

Median: $\frac{7 + 10}{2} = 8\frac{1}{2}$

Mean: $\frac{5 + 7 + 10 + 10}{4} = 8$

Mode: 10

Make a frequency table.

1. Abra asked her friends to name their favorite fruit. The responses are listed below.

pear	orange	apple	kiwi
orange	apple	grapes	apple
apple	orange	banana	grapes
apple	banana	grapes	apple
apple	pear	orange	apple
orange	banana	orange	orange
banana	orange	banana	apple
kiwi	apple		

Favorite Fruit		
Fruit	**Tally**	**Total**
pear		
grapes		
apple		
orange		
banana		
kiwi		

Use the data in exercise 1 for exercises 2–5.

2. Organize the data from least to greatest in the table at the right.

Favorite Fruit	
Kind	**Numbers**

3. How many friends did Abra question?

____

4. Which fruit was the favorite of the greatest number of Abra's friends? the least number?

____

5. Find the range, median, mean, and mode. ______________________

Find the range, median, mean, and mode for each set of data. Do your work on a separate sheet of paper.

6. 48, 45, 50, 52

____

7. 82°F, 85°F, 81°F, 88°F, 79°F

____

8. 23, 44, 28, 64, 32, 44, 38,

____

9. 120, 94, 78, 94, 88, 94, 97

____

10. \$22, \$9, \$54, \$36, \$38, \$29, \$36

____

11. \$135, \$120, \$144, \$136, \$95

____

Line Plots

Name ______________________

Date ______________________

The line plot shows the number of push-ups the members of the soccer team were able to do.

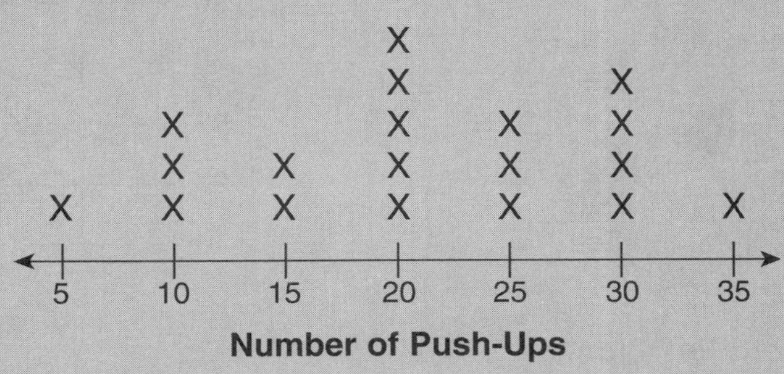

Use the line plot above to answer questions 1–5.

1. What is the mode of the data? ____________
2. What is the range of the data? ____________
3. Around which number do the data seem to cluster? ____________
4. How many team members could do 20 or more push-ups? ____________
5. How many team members could do fewer than 20 push-ups? ____________

Draw a line plot for each set of data. Find the range and mode.

6. Donovan's basketball scores: 14, 17, 20, 18, 16, 18, 17, 20, 15, 18, 16, 14, 16, 18, 15

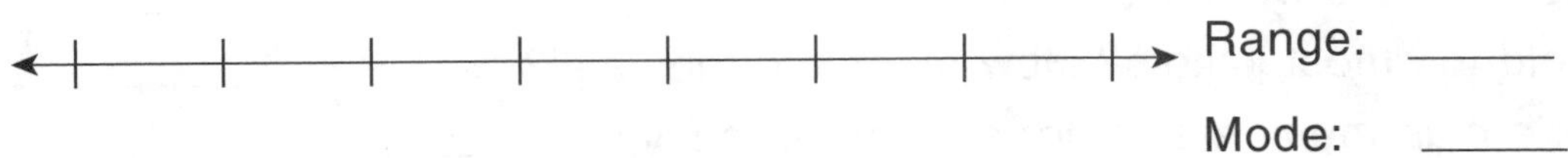

Range: ______

Mode: ______

7. Edie's math scores: 83, 93, 80, 85, 85, 87, 88, 90, 84

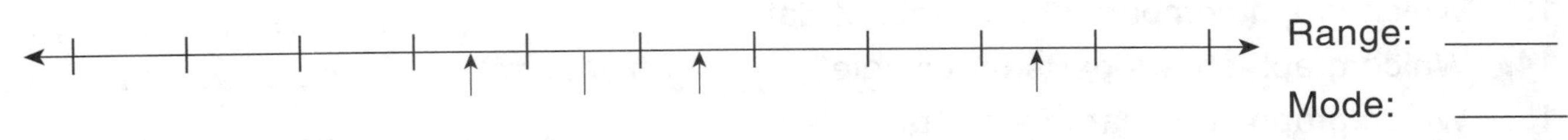

Range: ______

Mode: ______

Working with Graphs

Name ______________________________

Date ______________________________

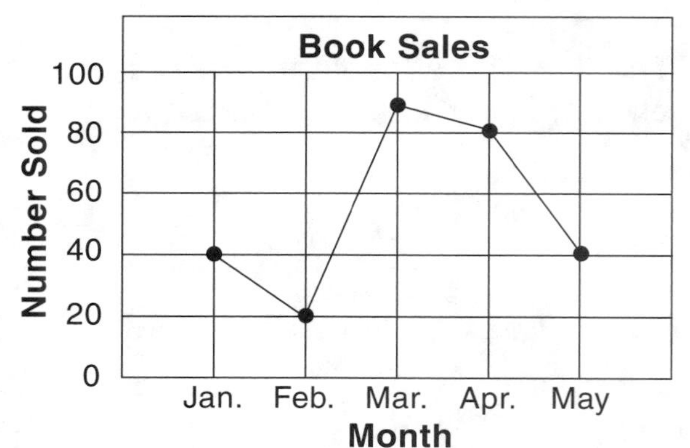

Tickets Sold For Picnic

Pony Club	
Little League	
Dance Club	
Key: = 10 tickets = 5 tickets	

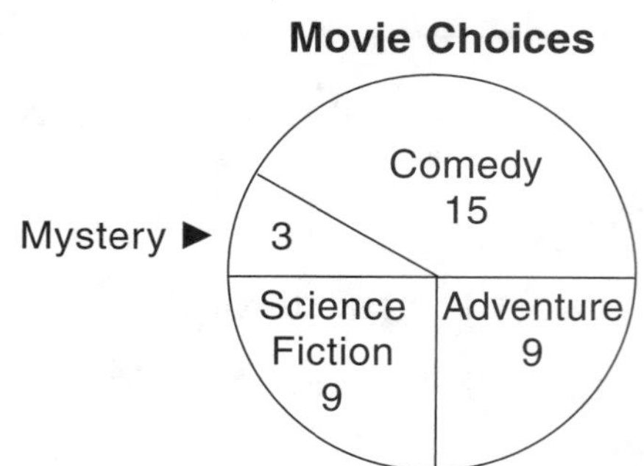

PROBLEM SOLVING Use the graphs above.

1. How many hours worked does each unit on the vertical scale represent? ______________
2. Who worked the greatest number of hours in one week? How many hours? ______________
3. How many hours did Ron work in one week? ______________
4. During which month were the greatest number of books sold? How many were sold? ______________
5. Between which two months was the decrease in book sales the greatest? ______________
6. Between which two months did the book sales increase? ______________
7. How many tickets does the symbol represent? ______________
8. How many tickets did the Dance Club sell? ______________
9. Which group sold the most tickets? How many did they sell? ______________
10. How many movie club members prefer adventure movies? ______________
11. What fractional part of the movie club prefers science fiction movies? ______________
12. Which type of movies is preferred by the fewest movie club members? ______________
13. Which graph compares three sets of data? ______________
14. Which graph shows parts of a whole? ______________
15. Which graphs compare four sets of data? ______________

 Use with Lesson 7-8, text pages 252–253.

Making Line Graphs

Name ______________________________

Date ______________________________

Use the table to complete the line graph. Then answer problems 2–4.

1.

Shoe Shack	
Day	**Pairs Sold**
Mon.	16
Tue.	32
Wed.	22
Thur.	28
Fri.	12

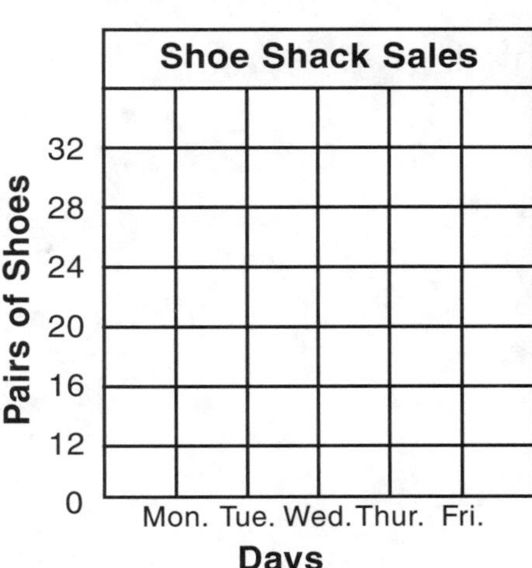

2. Between which two days was the difference in sales the least?

____

3. Between which two days was the increase in sales the most?

____

4. On the average, about how many pairs of shoes were sold daily?

____

Make a line graph for the data below.

5.

Average Monthly Temperature in Washington, DC (°F)	
Month	**Temperature**
Sept.	67
Oct.	55
Nov.	45
Dec.	35
Jan.	31
Feb.	34
Mar.	43

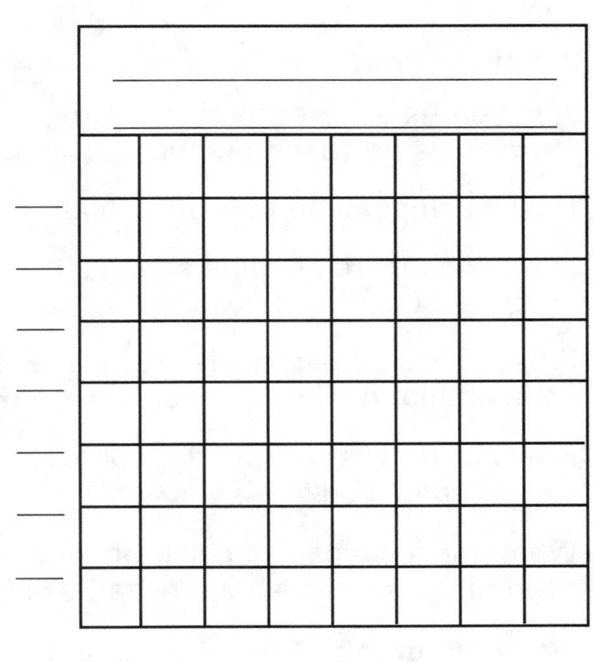

Use the graph you made in problem 5.

6. Which month is the warmest? the coldest?

____

7. About how many degrees difference is there between February and October?

____

8. About what is the average monthly temperature for these months

____

9. Between which consecutive months was there an increase of about 10°F?

____

Interpreting Circle Graphs

Name ______________________________

Date ______________________________

Use the circle graph at the right to answer problems 1–3.

Number of Cars Purchased

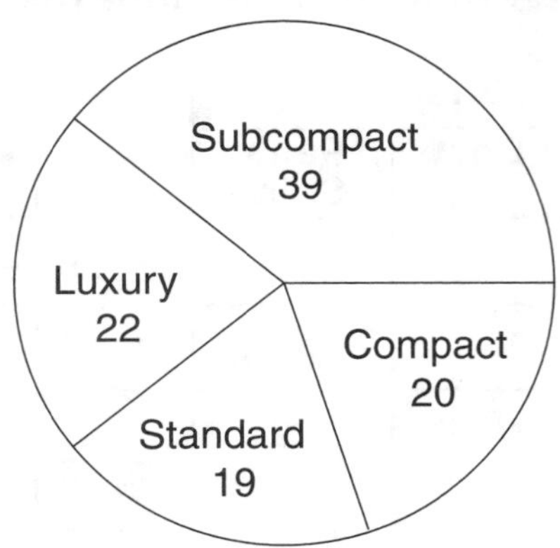

1. How many cars in all are represented by the graph? __________

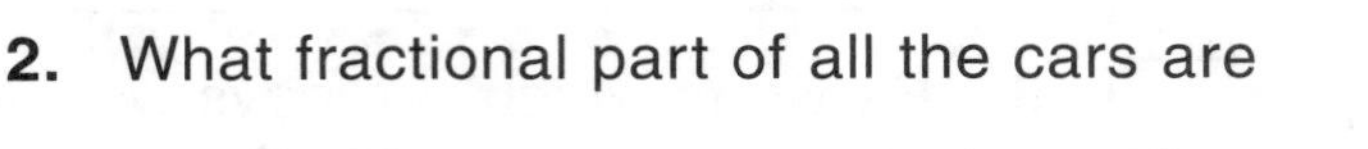

2. What fractional part of all the cars are

   compact? __________ luxury? __________

   standard? __________ subcompact? __________

3. Which two types of cars together equal the number of subcompact cars purchased?

   ____

Use the circle graph at the right to answer problems 4–9.

Hank's Budget

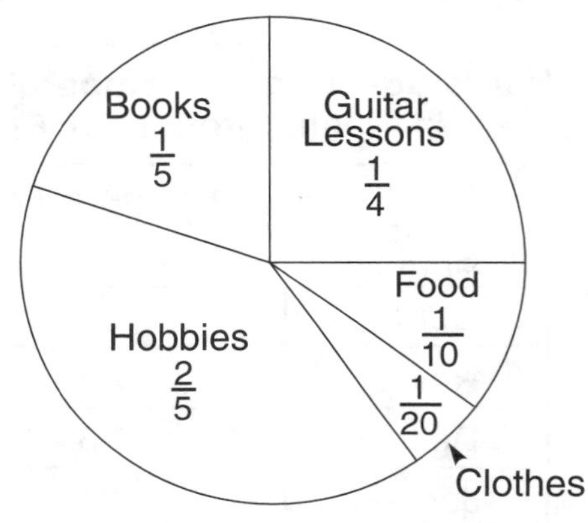

4. Hank earns $120 each month. How much does he spend on guitar lessons? ________
5. How much does he spend for food? ________
6. How much more does he spend for guitar lessons than books? ________
7. What fractional part of Hank's budget does he spend for books, clothes, and food? ________
8. How much does he spend for books, clothes, and food? ________
9. What fractional part of Hank's budget does he spend for his hobbies and guitar lessons? ________

Use the circle graph at the right to answer problems 10–12.

Favorite Sport

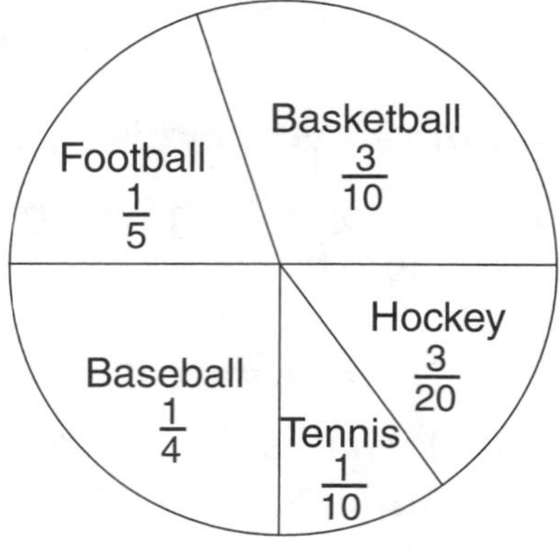

10. Which sport does the greatest number of students favor? __________
11. Which sport is the least favored? __________
12. If 200 students took part in the survey, how many chose:

    a. football? __________ **b.** baseball? __________

    c. hockey? __________ **d.** tennis? __________

    e. basketball? __________

 Use with Lesson 7-10, text pages 256–257.

Problem-Solving Strategy: Use a Model/Diagram

Name ______________________

Date ______________________

In a group of 25 students, $\frac{2}{5}$ of them play only basketball, 7 play only soccer, and 5 play neither sport. How many students play both basketball and soccer?

Draw a diagram.

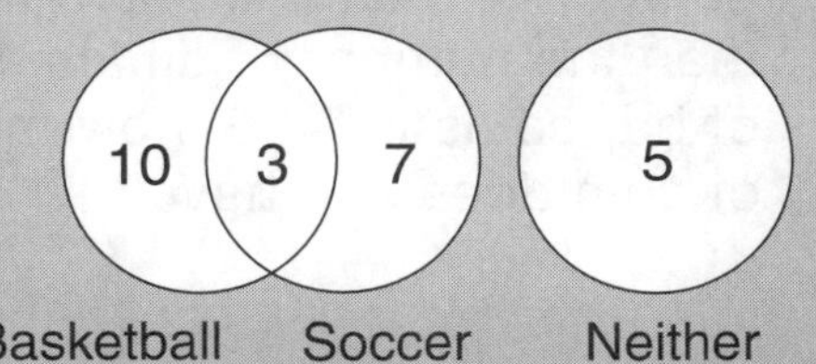

To find how many students just play basketball, multiply:

$\frac{2}{5} \times 25 = 10$

To find how many students play both sports, subtract:

$25 - 5 - 10 - 7 = 3$.

Three students play both basketball and soccer.

Solve. Do your work on a separate sheet of paper.

1. Of 18 students, $\frac{1}{3}$ can play guitar and piano, 6 can play only the guitar, and 4 can play neither instrument. How many students can play only the piano?

   ____

2. The History Club can go to Boston, Philadelphia, or Washington, DC. They can travel by plane or by train. Draw a tree diagram to find the number of travel choices they have.

   ____

3. Use the diagram below. What is the total number of students who play at least 1 of the 3 sports? the number of students who play only 1 of the 3 sports? the number of students who play all 3 sports?

   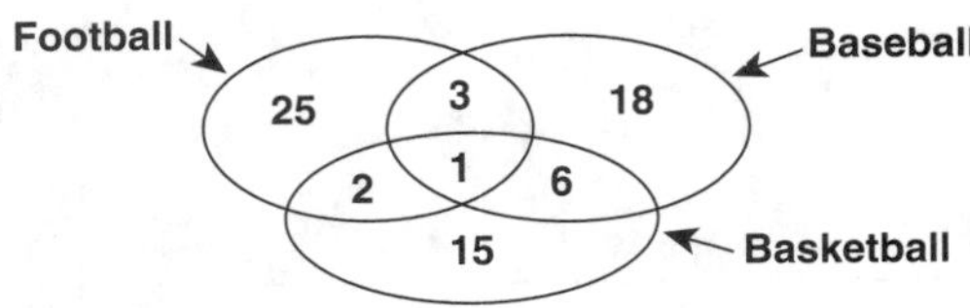

   ____

4. You toss a coin that can land heads or tails. Then you roll a number cube that has 2 red faces, 1 green face and 3 blue faces. Draw a tree diagram to show the possible outcomes. What is the probability that you will toss a head and roll a blue face?

   ____

5. There are 18 bottles of fruit juice on a shelf. Four bottles contain only apple juice. Nine bottles contain orange juice, and 7 contain grapefruit juice. How many bottles contain both orange and grapefruit juice?

   ____

6. Alice, Bruce, and Charlene each borrow a mystery book from the library. Alice, Doug, and Pat each borrow a biography. Charlene, Elaine, and Pat each borrow a travel book. How many different students borrow at least 2 books from the library?

   ____

Problem Solving: Review of Strategies

Name ______________________________

Date ______________________________

Solve. Do your work on a separate sheet of paper.

1. Eva has only quarters and dimes. The number of quarters she has is 1 more than the number of dimes. If the total value of her coins is $.95, how many of each kind of coin does she have?

____

2. Elsa collects baseball cards. She has collected an average of 1 card a week for each of the last 4 years. If she wants to put 8 cards on each page of an album, what is the least number of pages the album should have?

____

3. Wendy bought a long-sleeved blouse for $24.99 and a pair of jeans for $34.56. She gave the clerk 3 twenty-dollar bills. How much change did she receive? What possible combination of coins and bills could she receive without receiving any pennies?

____

4. Kevin has nickels, dimes, and quarters. He has twice as many dimes as nickels. The number of quarters is 1 fewer than the number of dimes. Kevin has 9 coins with a total value of $1.25. How many of each kind of coin does he have?

____

5. Alvin, Philip, and Alex have each invited a sister of each of the other boys to the Junior Prom. No girl is going to the prom with her brother. Alice is not Alvin's sister and is not going to the prom with Dulce's brother. Lauren is Philip's sister. Name the girl that each boy has invited to the prom.

____

6. Of the 100 ninth-grade students in East High School, 30 take French and biology, 10 take French but not biology, and 12 take neither French nor biology. How many ninth-grade students of East High School take only biology?

____

7. Avery wants to buy 2 dozen ball-point pens. Bee Brand Pens are priced at $5.52 per dozen. Bear Brand Pens are priced at $.49 each. Assuming both brands are of equal quality, which brand should Avery buy? How much would he save?

____

8. Roy has 100 baseball cards. He arranges them in 3 stacks so that the second stack has twice as many cards as the first and the third stack has twice as many cards as the second. What is the largest number of cards that he places in each stack? How many cards are left over?

____

Use after Chapter 7.

Measuring and Drawing Angles

Name ____________________

Date ____________________

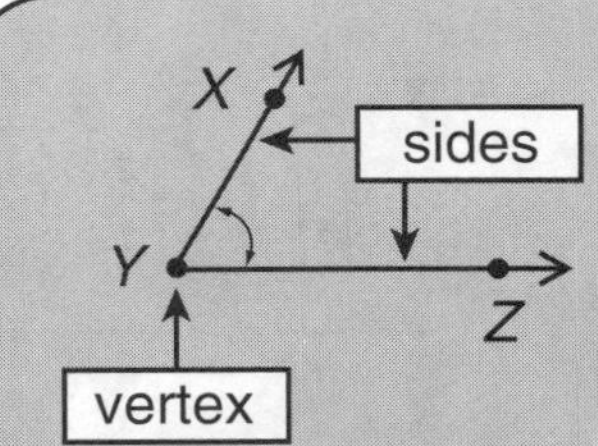

Name: $\angle Y$ or $\angle XYZ$ or $\angle ZYX$
Sides: $\overrightarrow{YX}$, $\overrightarrow{YZ}$
Vertex: Y
$\angle Y$ measures 60°.

Name the angle, its sides, and its vertex.

1. 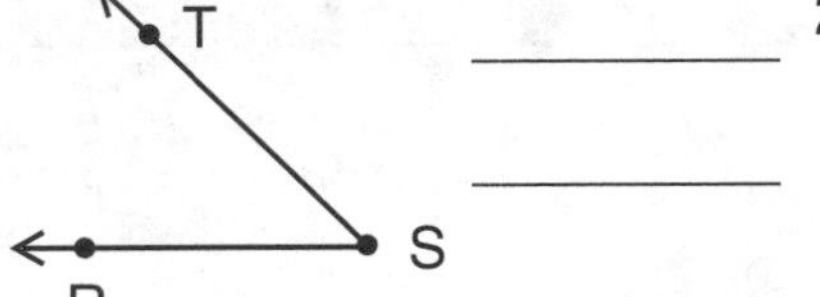

2. 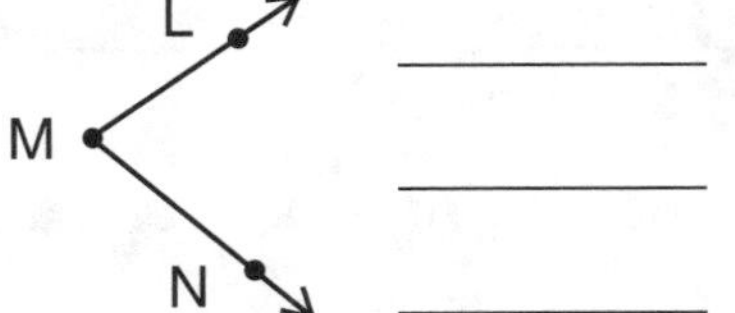

3. 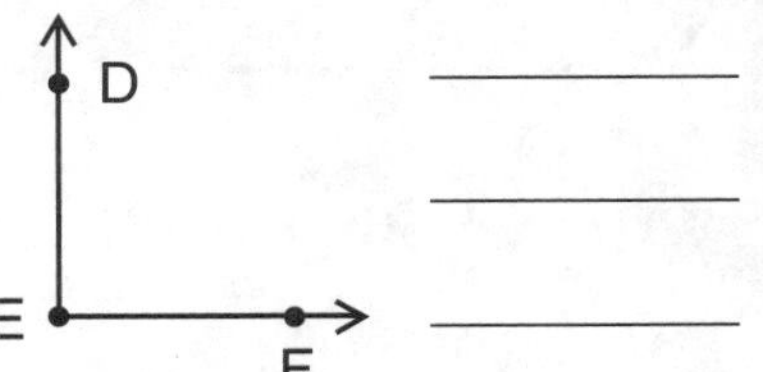

Use a protractor to find the measure of each angle.

4. 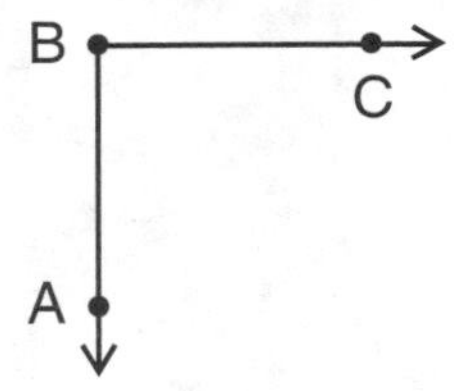

5. 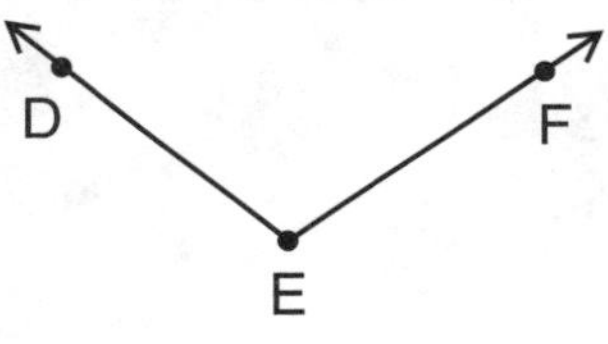

6. 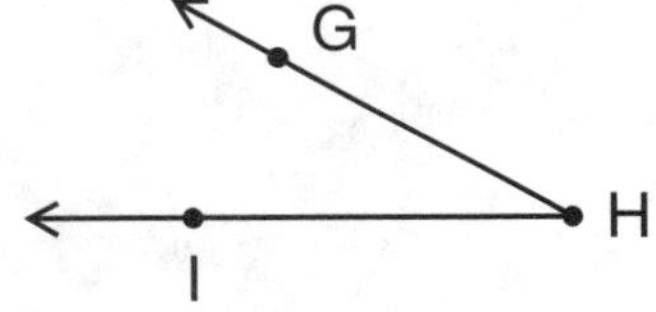

7. 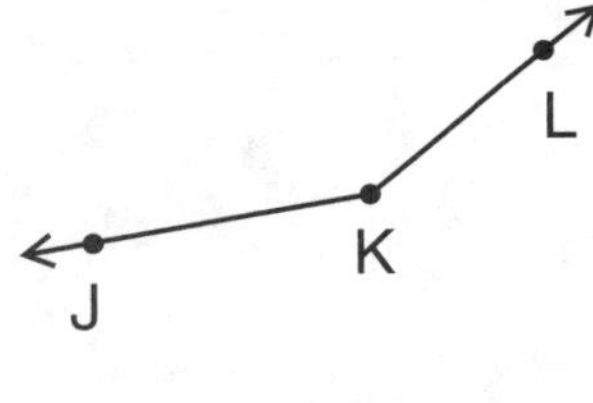

8. 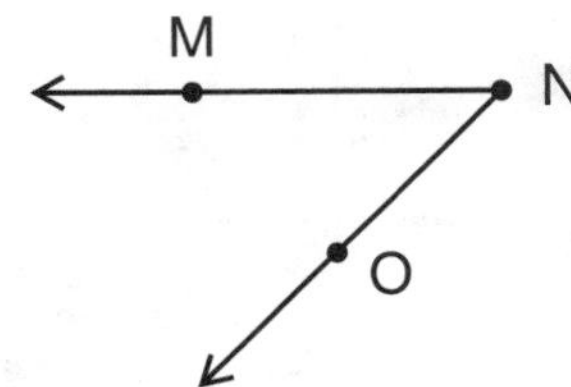

9. 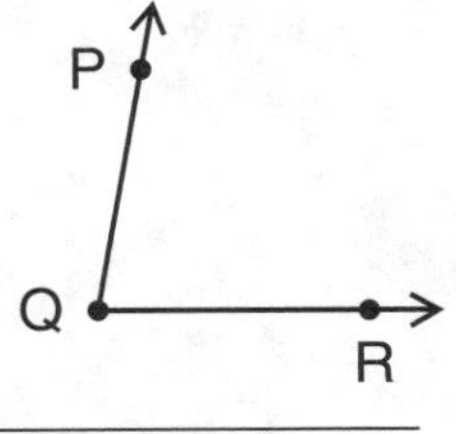

Use a protractor to draw each angle. Then label each angle.

10. 85° **11.** 130° **12.** 100°

Use with Lesson 8-1, text pages 268–269.

Identifying Angles

Name ______________________

Date ______________________

right angle

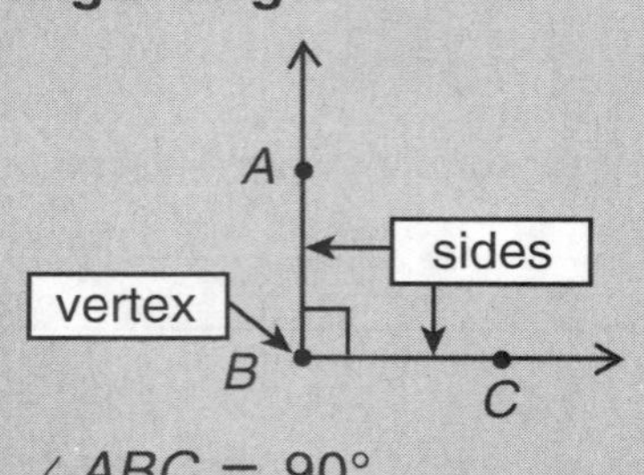

$\angle ABC = 90°$

acute angle

$\angle DEF < 90°$

obtuse angle

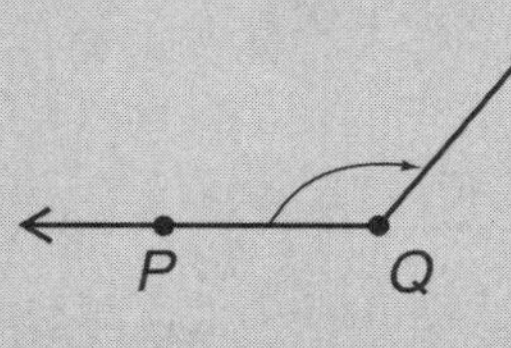

$\angle PQR > 90°$

straight angle

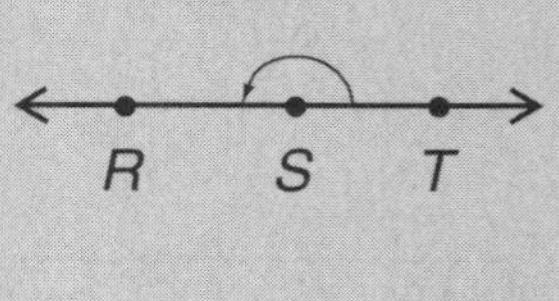

$\angle RST = 180°$

Write whether each angle is *acute*, *right*, *obtuse*, or *straight*.

1. 130° ________ **2.** 15° ________ **3.** 88° ________ **4.** 90° ________

5. 7° ________ **6.** 94° ________ **7.** 174° ________ **8.** 180° ________

Classify each angle as *acute*, *right*, *obtuse*, or *straight*.

9.

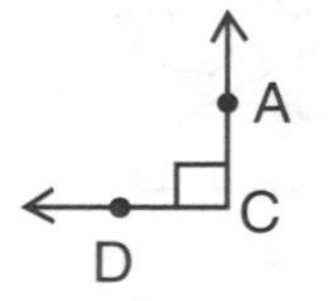

____

10.

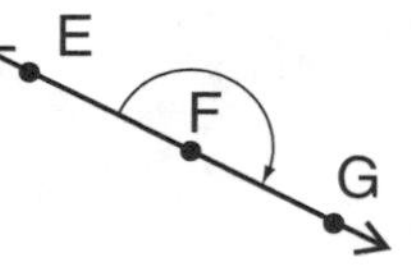

____

11.

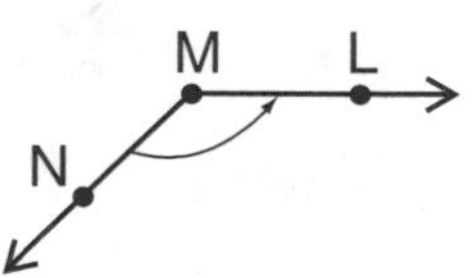

____

12.

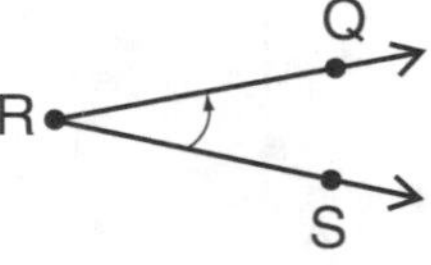

____

13.

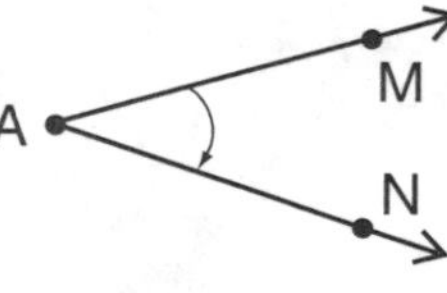

____

14.

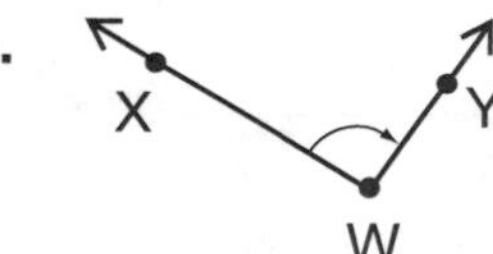

____

15.

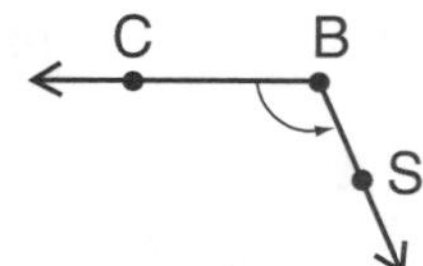

____

Use the figure to complete exercises 16–18.

16. Name an acute angle. ______________________

17. Name an obtuse angle. ______________________

18. Name a right angle. ______________________

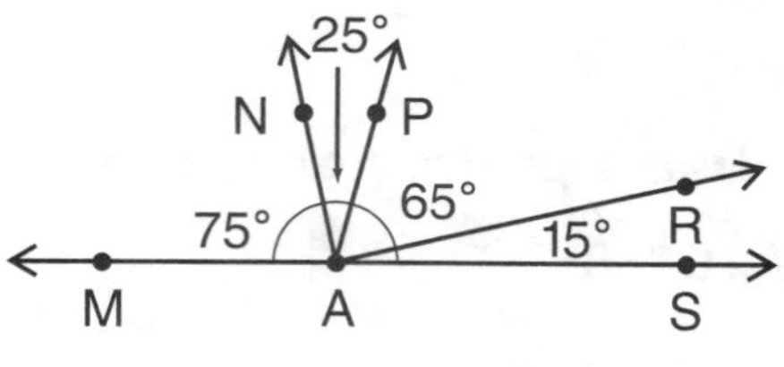

Are the lines perpendicular? Write *Yes* of *No*.
Use a protractor to check.

19.

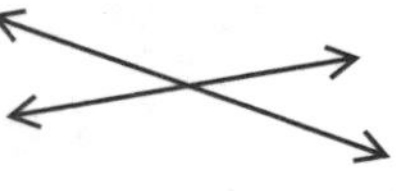

____

20.

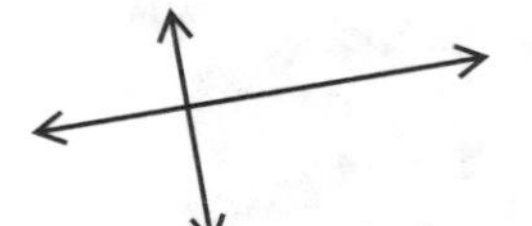

____

21.

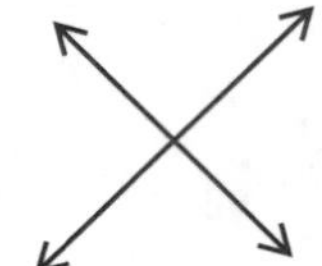

____

22.

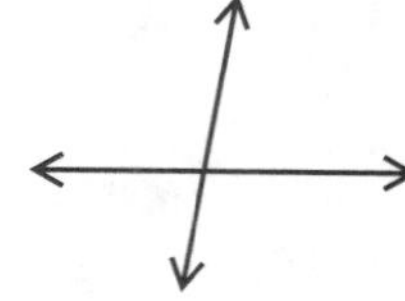

____

 Use with Lesson 8-2, text pages 270–271.

Polygons

Name ______________________

Date ______________________

A polygon is a closed plane figure made up of line segments that meet at vertices but do not cross.

Polygons

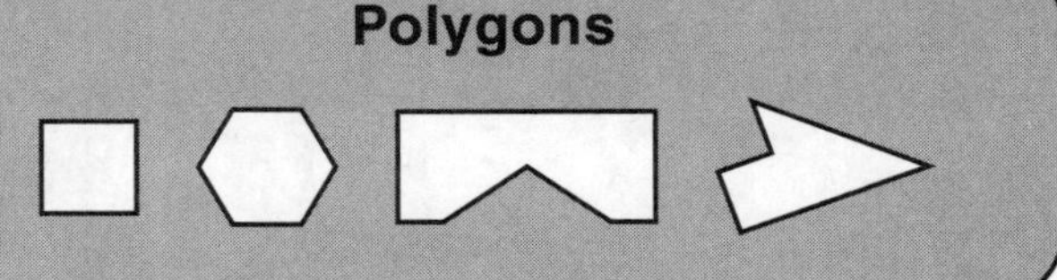

Write the number of sides and vertices.

1. 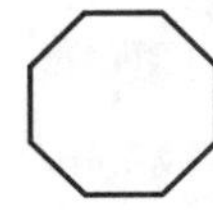

Sides: ________
Vertices: ________

2. 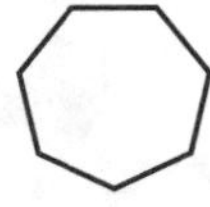

Sides: ________
Vertices: ________

3. 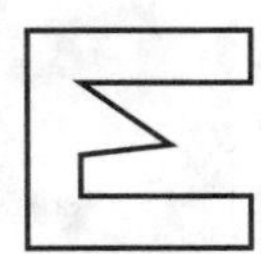

Sides: ________
Vertices: ________

4. 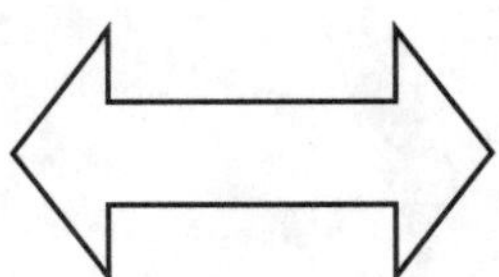

Sides: ________
Vertices: ________

Complete the table. Then answer the question.

	Name of Polygon	Number of Sides	Number of Angles
5.	**Heptagon**	7	
6.	**Octagon**		
7.	**Nonagon**	9	
8.	**Decagon**		10

9. What can you say about the number of sides and the number of angles for each polygon? ______________________

Draw four different polygons. Write the number of the sides and angles.

10. 

Sides: ________
Angles: ________

11.

Sides: ________
Angles: ________

12.

Sides: ________
Angles: ________

13.

Sides: ________
Angles: ________

 Use with Lesson 8-3, text pages 272–273.

Congruent Figures

Name ______________________

Date ______________________

Congruent Polygons

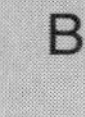

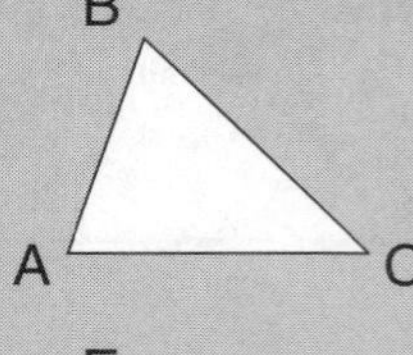

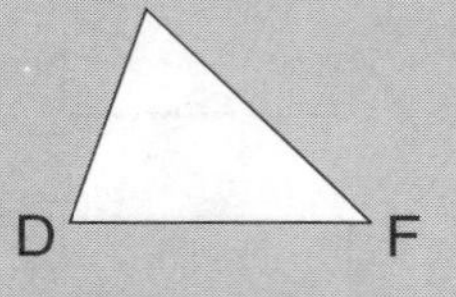

$\triangle ABC \cong \triangle DEF$

Corresponding Sides:

$\overline{AB} \cong \overline{DE}$
$\overline{AC} \cong \overline{DF}$
$BC \cong EF$

Corresponding Angles:

$\angle A \cong \angle D$
$\angle B \cong \angle E$
$\angle C \cong \angle F$

Similar Polygons

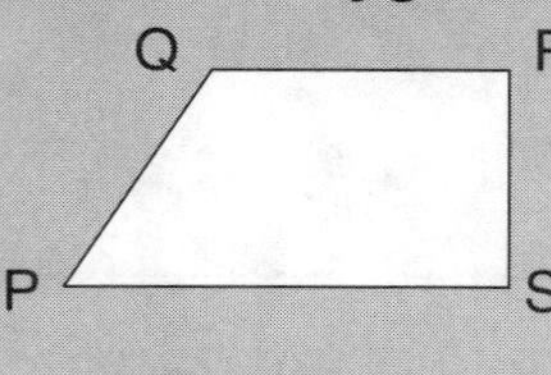

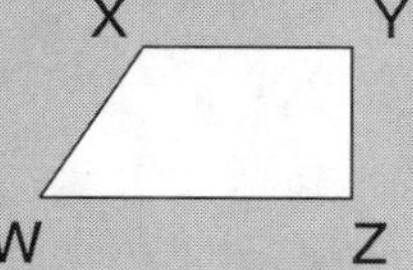

Corresponding Angles:

$\angle P \cong \angle W$
$\angle Q \cong \angle X$
$\angle R \cong \angle Y$
$\angle S \cong \angle Z$

Quadrilateral *PQRS* ~ Quadrilateral *WXYZ*

Are the figures congruent? Write *Yes* or *No*.

1.

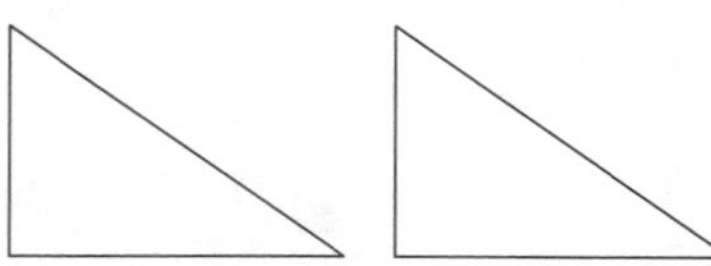

____

2.

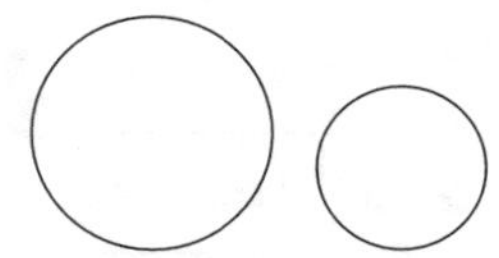

____

3.

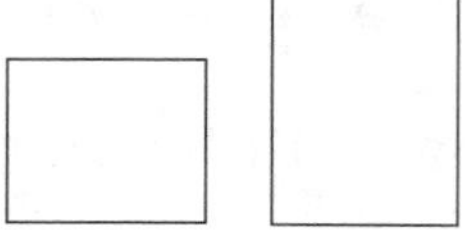

____

Complete. Use the figures at the right.

4. $\overline{LK} \cong$ __________ **5.** $\overline{LM} \cong$ __________

6. $\overline{KN} \cong$ __________ **7.** $\overline{MN} \cong$ __________

8. $\angle K \cong$ __________ **9.** $\angle M \cong$ __________

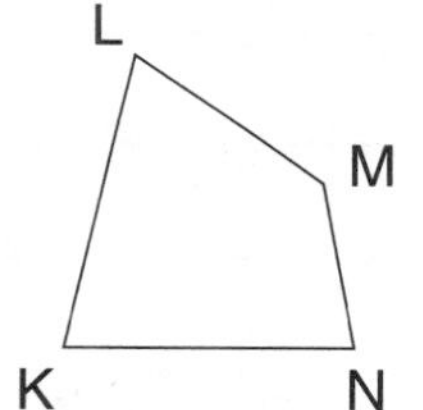

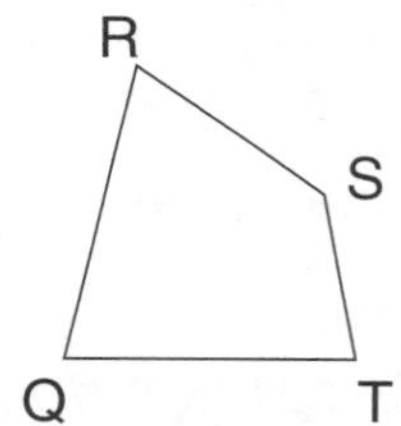

Quadrilateral *KLMN* $\cong$ Quadrilateral *QRST*

Are the figures similar? Write *Yes* or *No*.

10.

____

11.

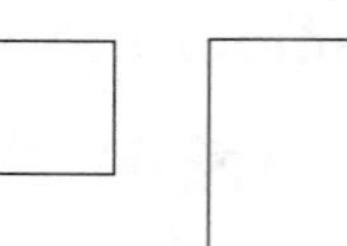

____

12.

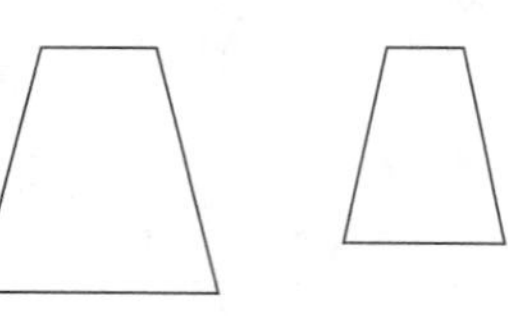

____

Use the symbol $\cong$ to identify corresponding angles.

13.

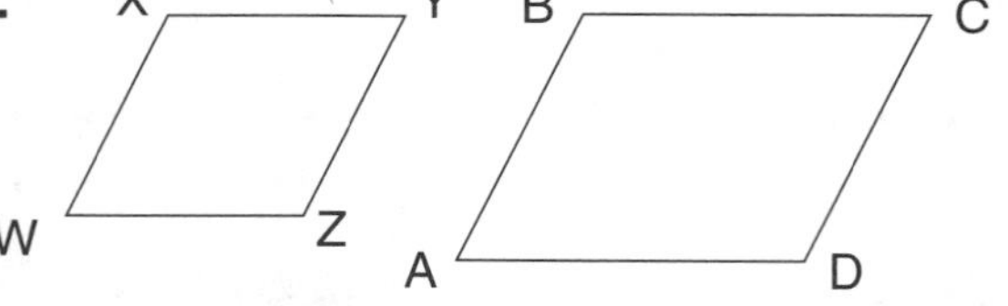

Quadrilateral *WXYZ* ~ Quadrilateral *ABCD*

____

14.

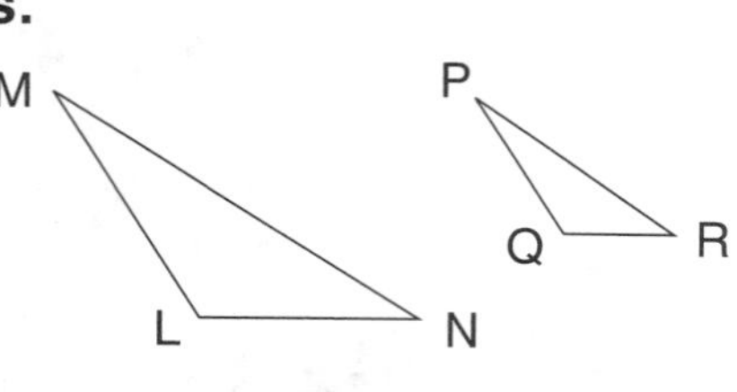

$\triangle MLN \sim \triangle PQR$

____

Use with Lesson 8-4, text pages 274–275.

Triangles

Name ______________________

Date ______________________

Triangles may be classified by the length of their sides or by the measures of their angles.

scalene

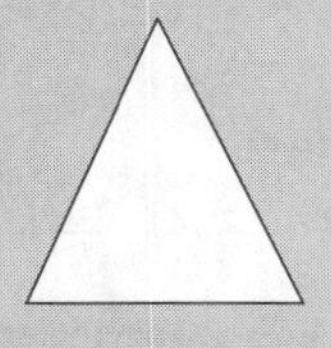
isosceles

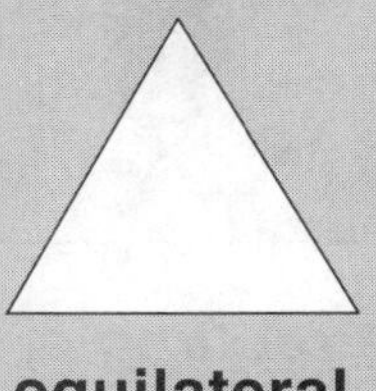
equilateral

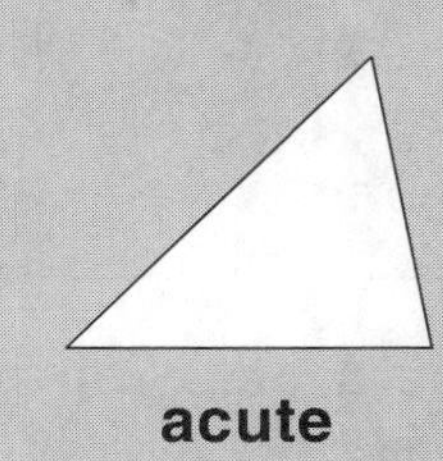
acute

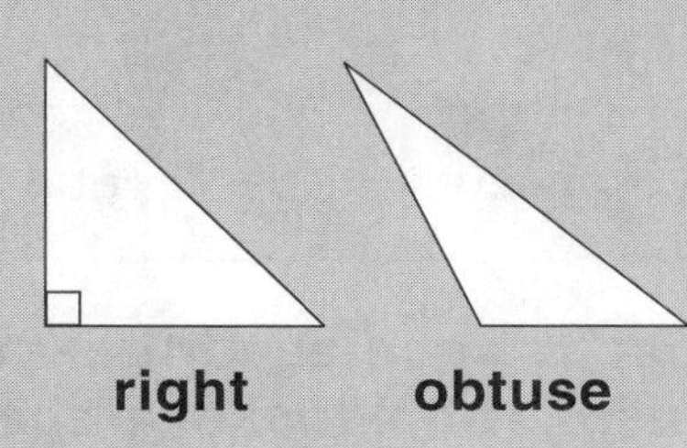
right obtuse

Classify each triangle as *scalene, isosceles,* or *equilateral.*

1.  ______________

2. 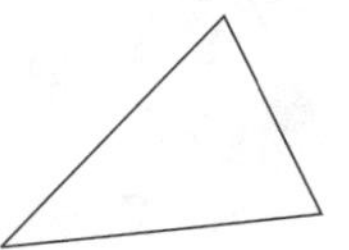 ______________

3. 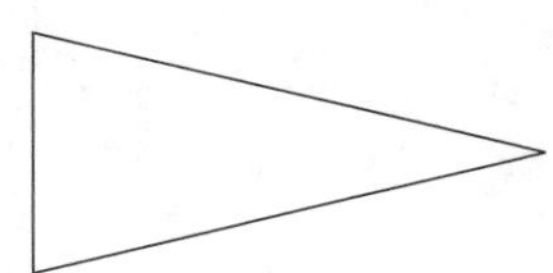 ______________

4. 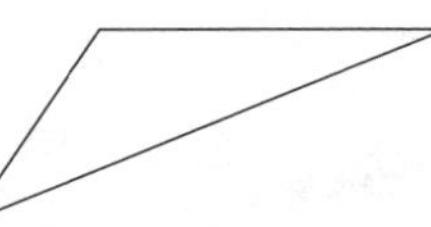 ______________

5. 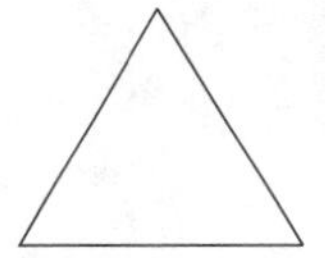 ______________

6. 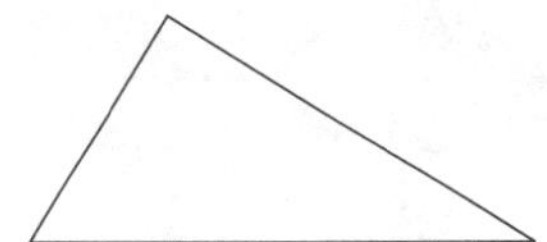 ______________

7. 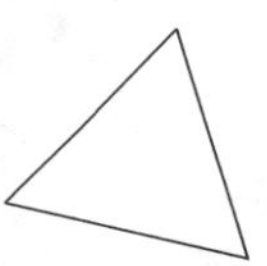 ______________

8. 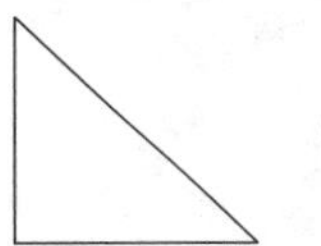 ______________

Classify each triangle as *acute, right,* or *obtuse*.

9. 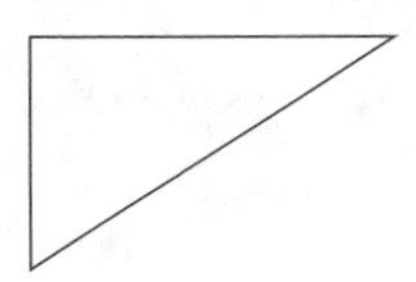 ______________

10. 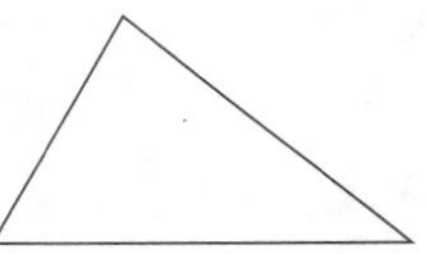 ______________

11. 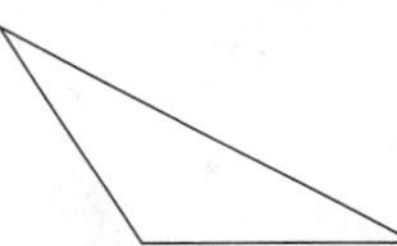 ______________

12. 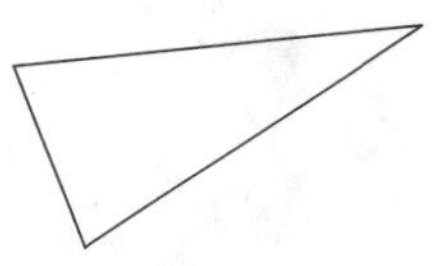 ______________

13. 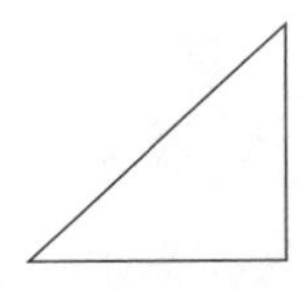 ______________

14. 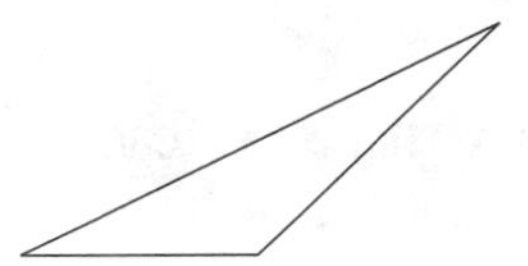 ______________

15. 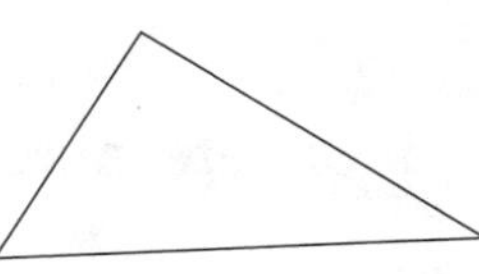 ______________

16. 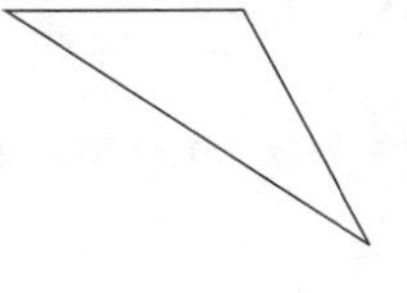 ______________

Is it possible for each triangle to have the given angle measures? Write *Yes* or *No.*

17.
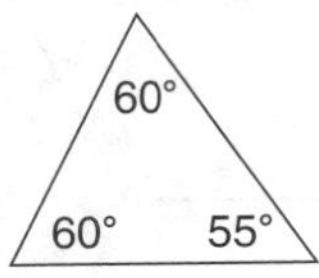

____

18.
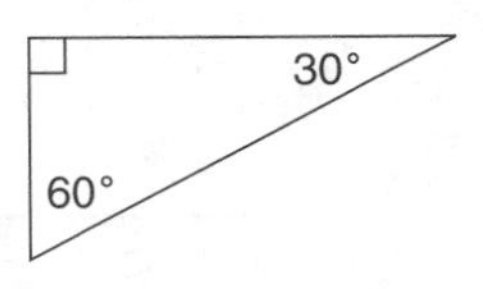

____

19.
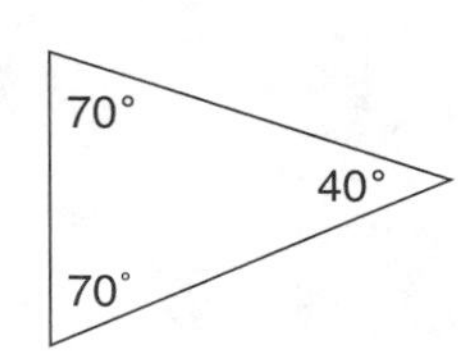

____

20.
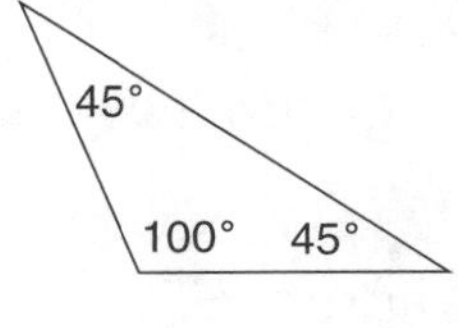

____

Use with Lesson 8-5, text pages 276–277.

Quadrilaterals

Name ______________________________

Date ______________________________

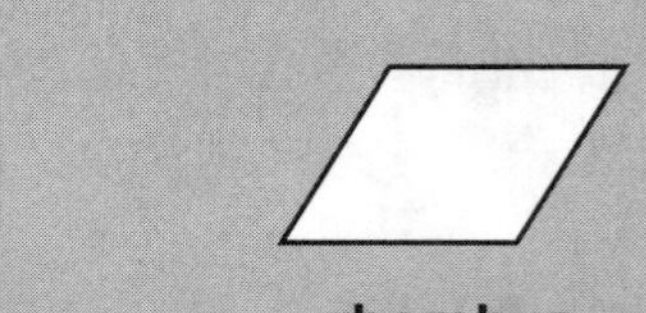

Classify each quadrilateral as a *parallelogram, rectangle, square, rhombus,* or *trapezoid.*

1. 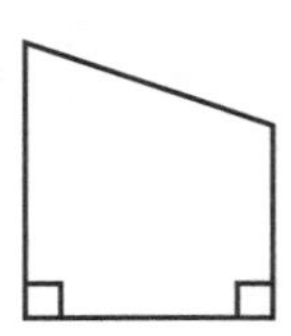 ______________

2. 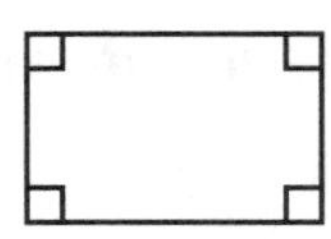 ______________

3. 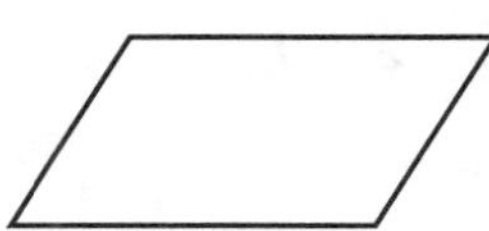 ______________

4. 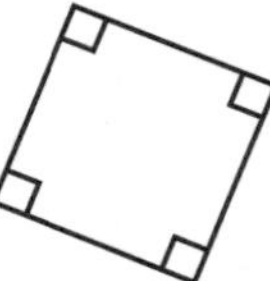 ______________

5. 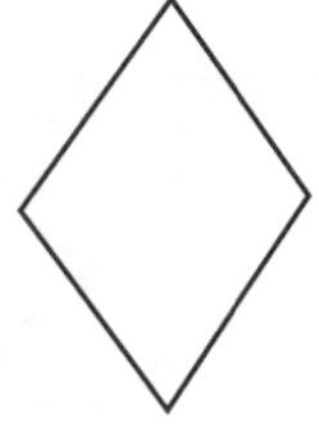 ______________

6. 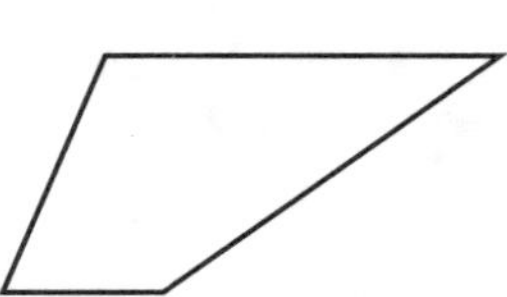 ______________

7. 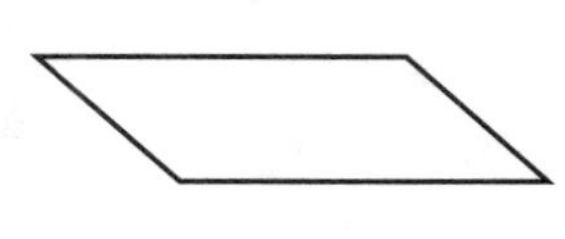 ______________

8. 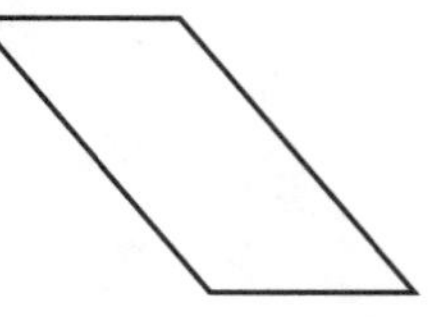 ______________

Draw all the diagonals for each figure. Then write the number of diagonals.

9. 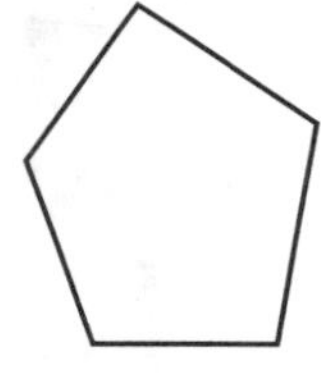 ______________

10. 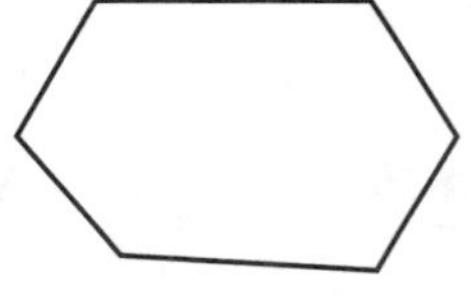 ______________

11.  ______________

Write the name of each quadrilateral. Use the figure below.

12. figure *ACEH* ______________

13. figure *BDJI* ______________

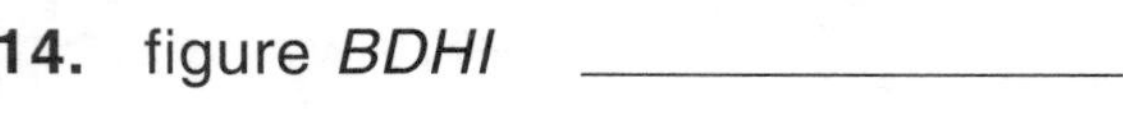

14. figure *BDHI* ______________

15. figure *EFGH* ______________

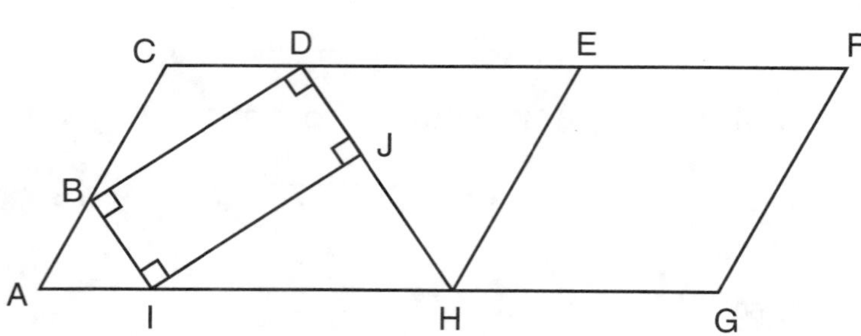

 Use with Lesson 8-6, text pages 278–279.

Perimeter of a Polygon

Name ______________________

Date ______________________

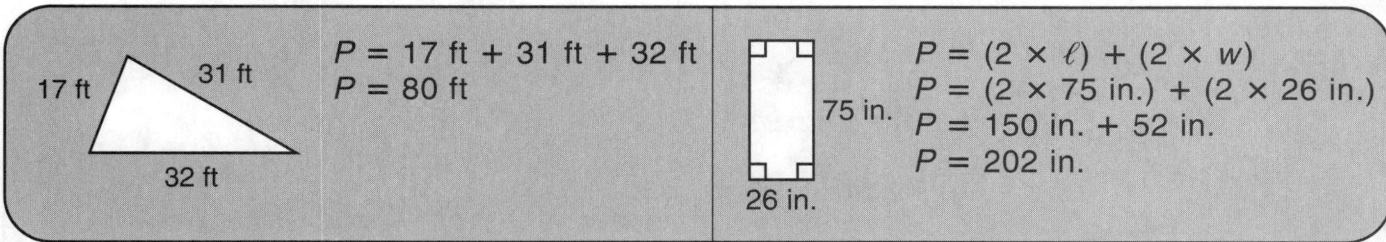

Find the perimeter of each polygon.

1.

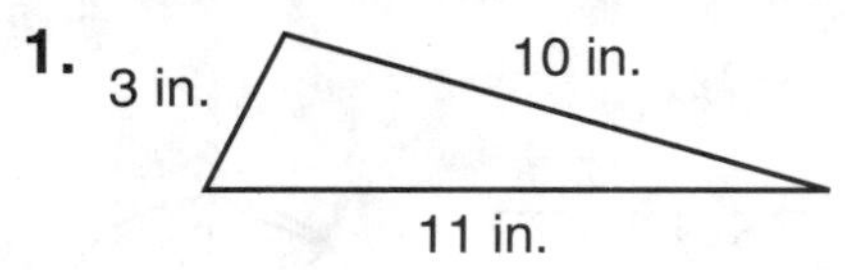

____

2.

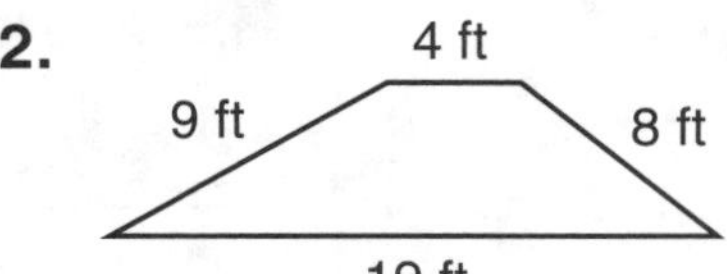

____

3.

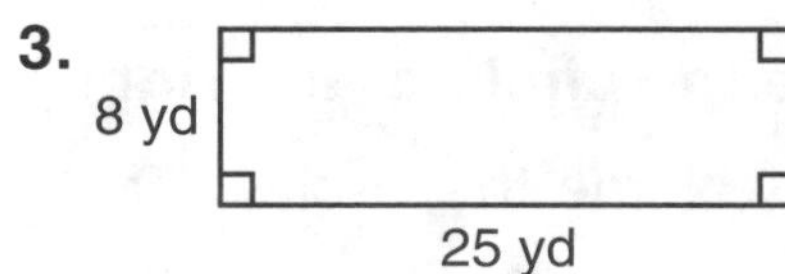

____

4.

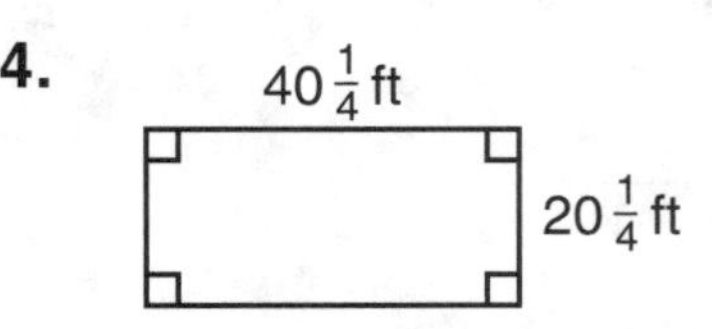

____

5.

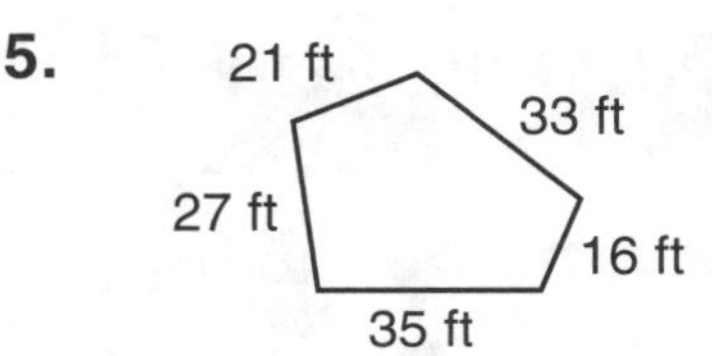

____

6.

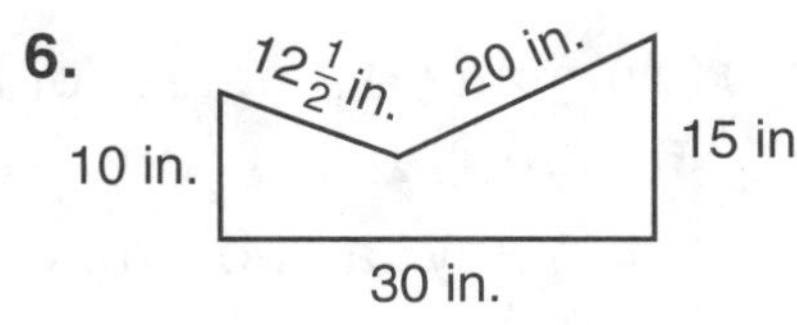

____

Complete.

	Polygon	Length of One Side	Perimeter
7.	square	14 in.	
8.	regular pentagon	6 ft	
9.	equilateral triangle	5 in.	
10.	regular hexagon		18 m
11.	regular octagon	7 cm	
12.	regular decagon		90 yd
13.	square	12 mm	

PROBLEM SOLVING You may draw a diagram.

14. A rectangular garden is 23 ft wide and 35 ft long. How many feet of fencing are needed to enclose the garden? ______________________

15. The sides of a quadrilateral have lengths of 18 in., 24 in., 35 in., and 19 in. Find the perimeter. ______________________

16. Each side of a square park measures 160 yd. If Gina walks twice around the outside of the park, how many yards does she walk? ______________________

 Use with Lessons 8-7, 8-8, text pages 280–283.

Circles

Name ______________________

Date ______________________

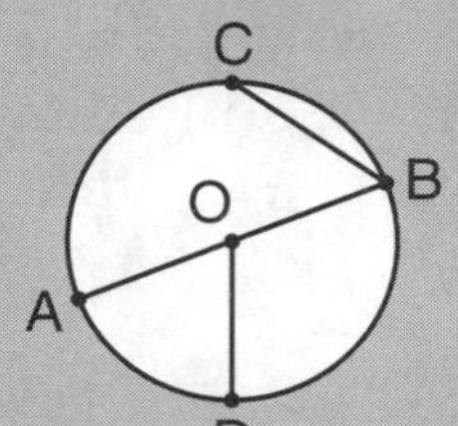

Center: *O*

Diameter: $\overline{AB}$

Radii: $\overline{OA}$, $\overline{OB}$, $\overline{OD}$

Chord: $\overline{BC}$

A circle is named by its center.

Use the circle at the right.

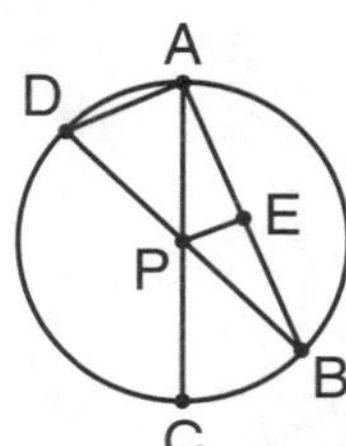

1. Name the circle. ______________
2. Name 4 points on the circle. ______________
3. How many chords of the circle are shown? Name them. ______________
4. How many diameters of the circle are shown? Name them. ______________
5. How many radii of the circle are shown? Name them. ______________

Complete each table.

	Diameter	Radius
6.		8 in.
7.	58 cm	
8.		$6\frac{1}{4}$ ft

	Diameter	Radius
9.	14 m	
10.		12 in.
11.	136 yd	

Use your compass to construct a circle on a separate sheet of paper. Then do the following:

12. Label the center point, C.
13. Draw radius $\overline{CD}$.
14. Draw chord $\overline{AE}$.
15. Draw diameter $\overline{ST}$.

PROBLEM SOLVING

16. The radius of a circular track is 18 m. Find the diameter. ______________
17. The diameter of a circular rug is 11 ft. What is the radius? ______________
18. Jill's bicycle wheel has a radius $23\frac{3}{5}$ cm. What is its diameter? ______________

Estimating Circumference

Name ______________________

Date ______________________

Estimate the circumference of each circle.

Diameter is given.

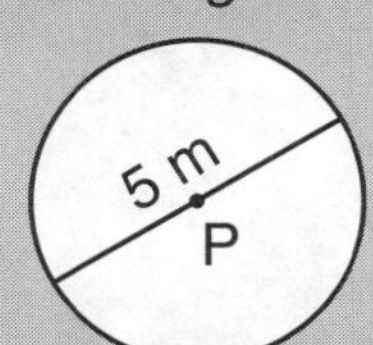

$C \approx 3 \times d$

$\approx 3 \times 5$ m

≈ 15 m

The circumference of circle *P* is about 15 m.

Radius is given.

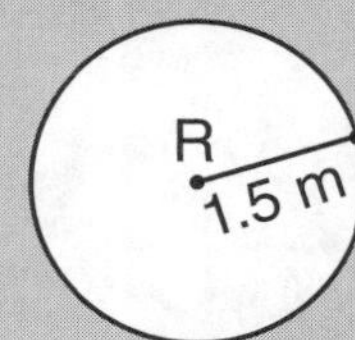

$C \approx 3 \times 2 \times r$

$C \approx 6 \times r$

$\approx 6 \times 1.5$ m

≈ 9 m

The circumference of circle *R* is about 9 m.

Estimate the circumference of each circle.

1. 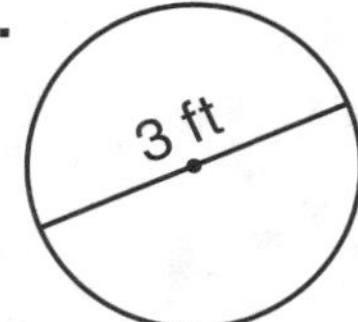

____

2. 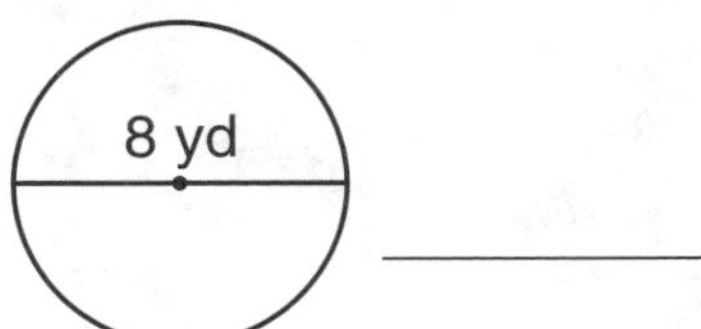

____

3. 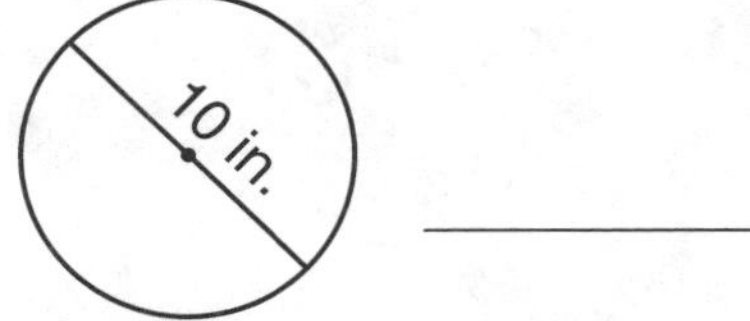

____

4. 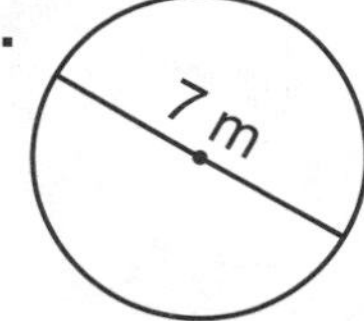

____

5. 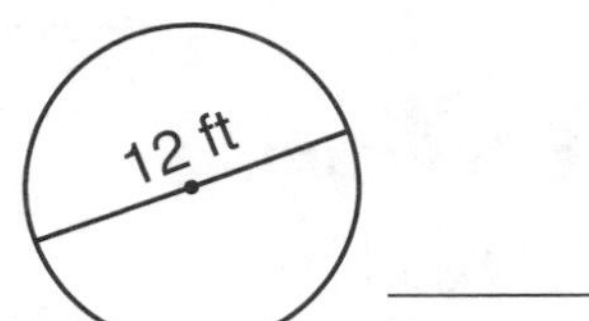

____

6. 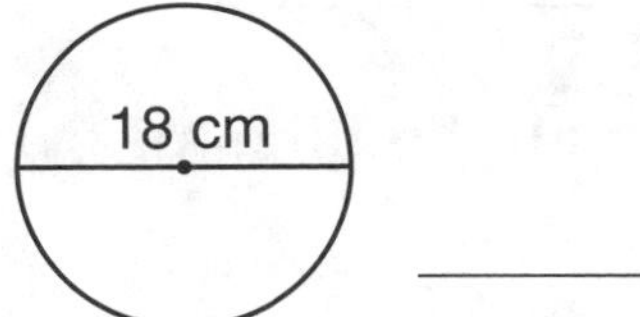

____

7. 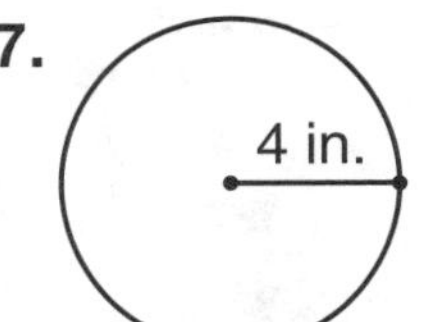

____

8. 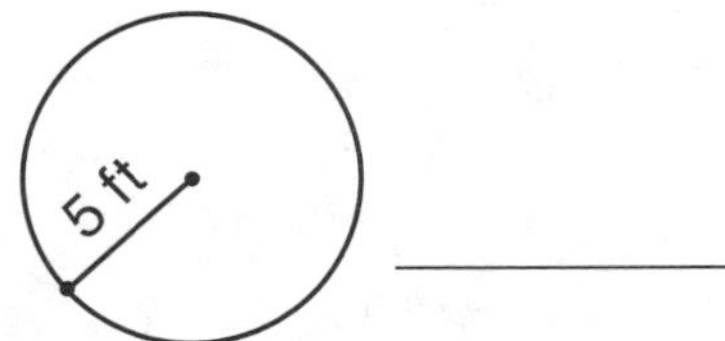

____

9. 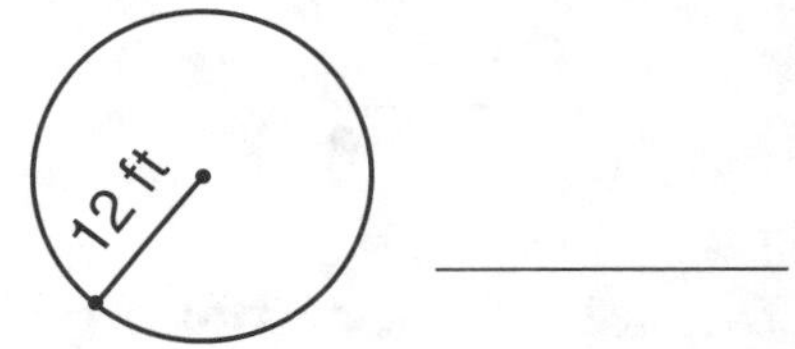

____

10. 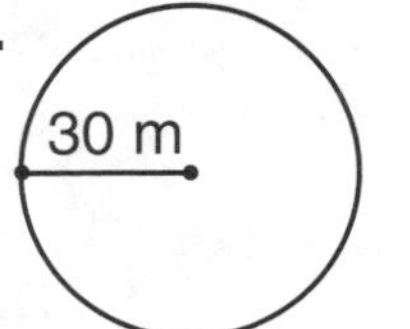

____

11. 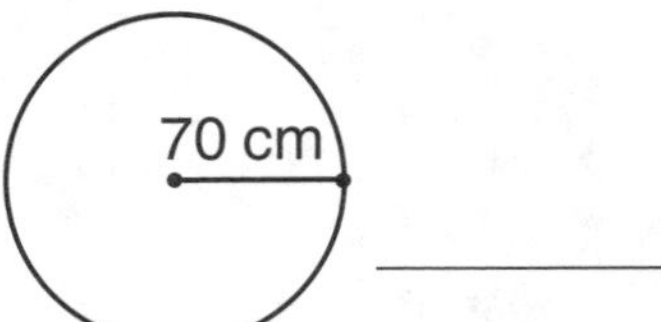

____

12. 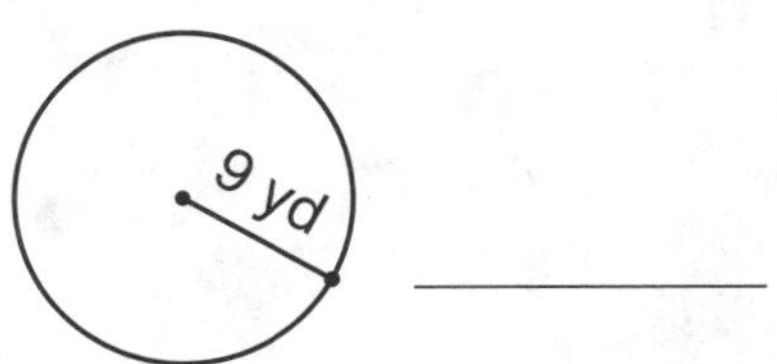

____

Estimate the circumference of each circle. Circle the letter of the correct answer.

13. 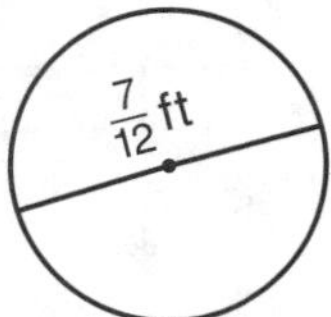

a. $1\frac{3}{4}$ ft

b. $1\frac{5}{12}$ ft

c. $1\frac{1}{2}$ ft

d. 3 ft

14. 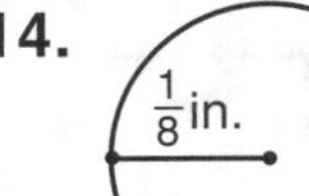

a. $\frac{1}{2}$ in.

b. $\frac{3}{8}$ in.

c. $\frac{3}{4}$ in.

d. 1 in.

Lines of Symmetry

Name ______________________

Date ______________________

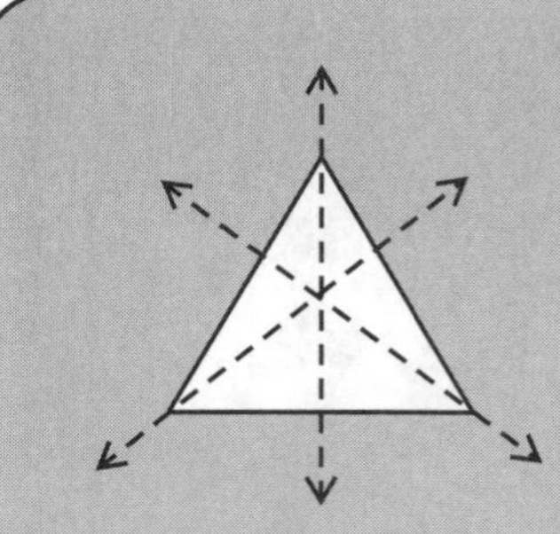

3 lines of symmetry

A rectangle has half-turn symmetry.

after half-turn

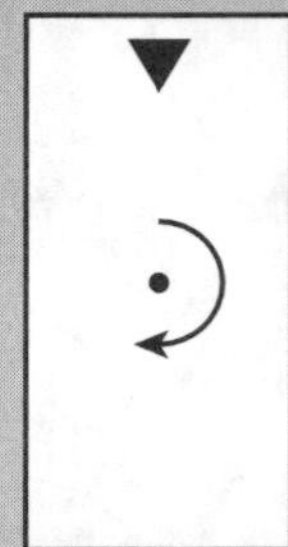

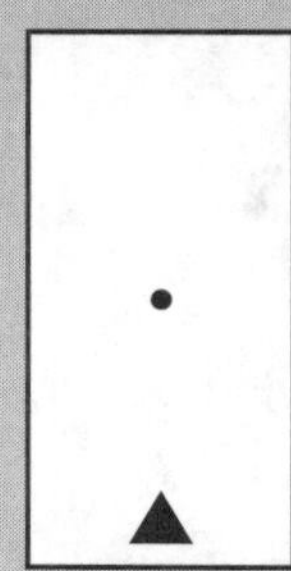

Is the dashed line a line of symmetry? Write *Yes* or *No*.

1. 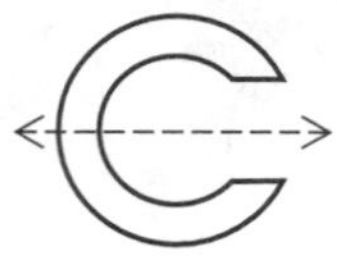 __________

2. 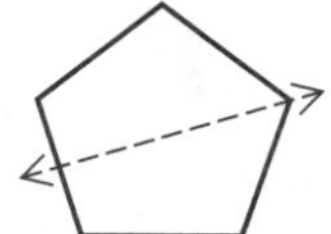 __________

3. 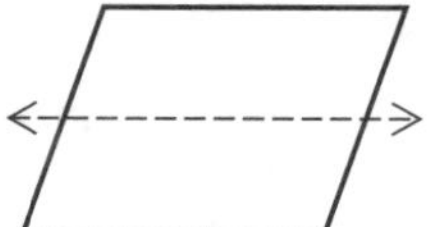 __________

4. 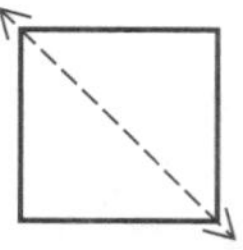 __________

Draw all the lines of symmetry for each figure.

5. 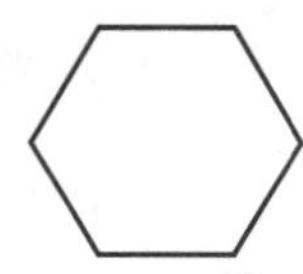

6. 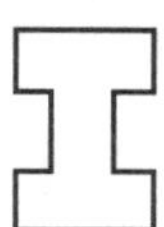

7. 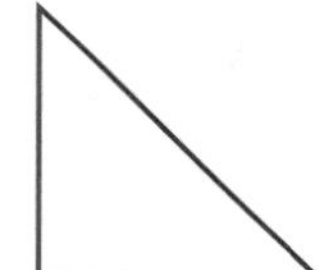

8. 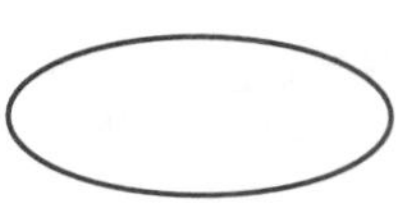

Does the figure have half-turn symmetry? Write *Yes* or *No*.

9. 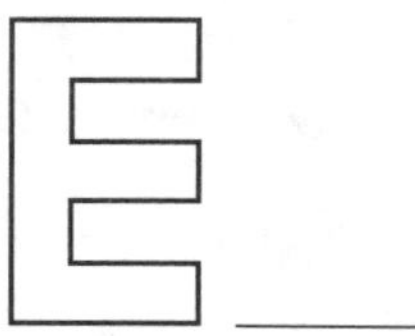 ______

10.  ______

11. 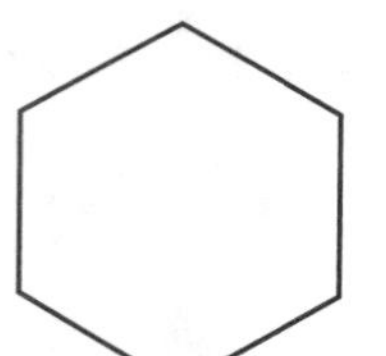 ______

12. 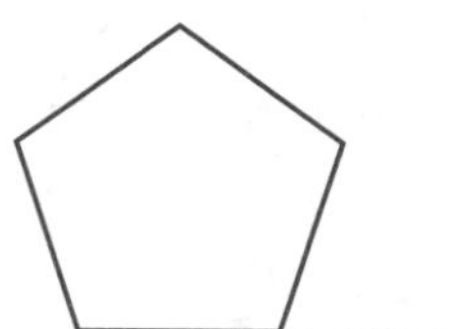 ______

Complete the figure so that the dashed line is the line of symmetry.

13. 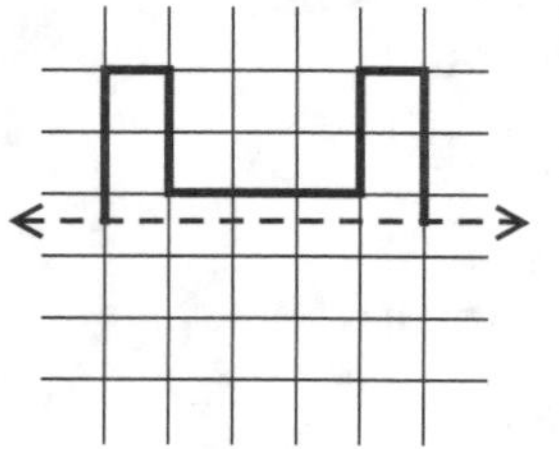

14. 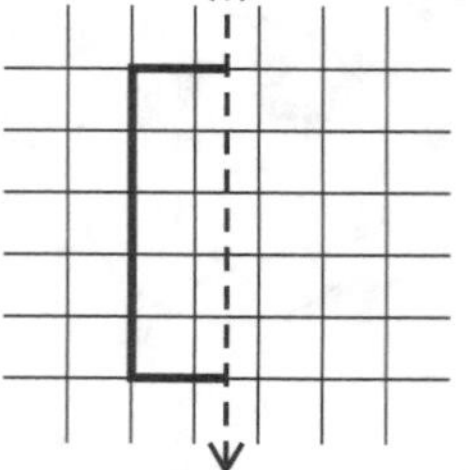

15. 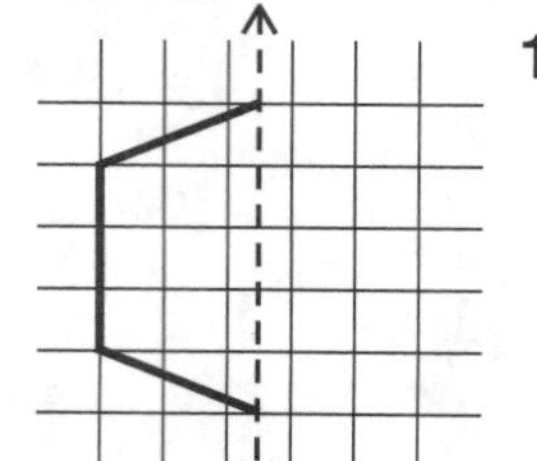

16. 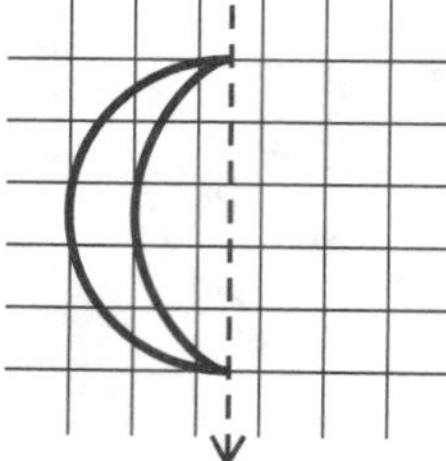

Use with Lesson 8-11, text pages 288–289.

Transformations

Name ____________________

Date ____________________

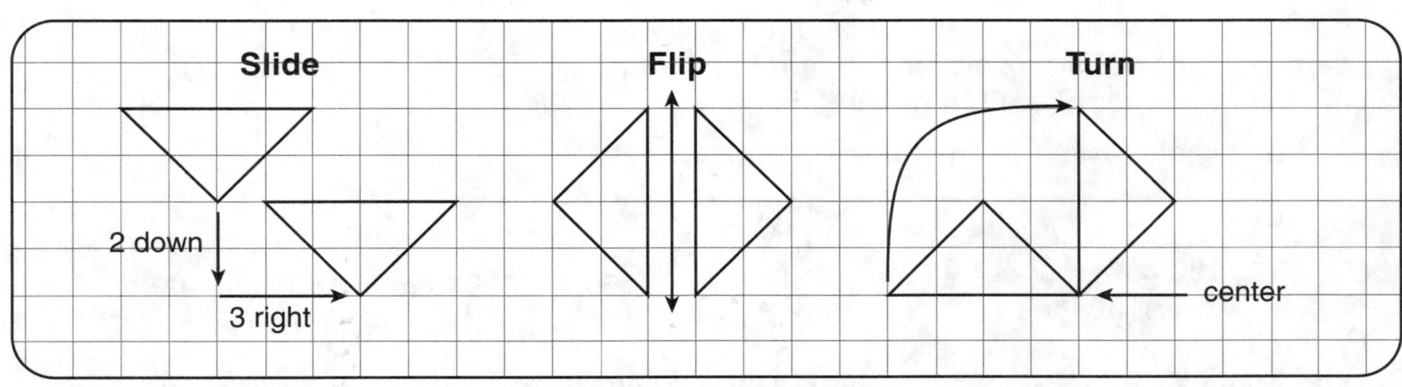

Is figure B a result of moving figure A? Write *Yes* or *No*.

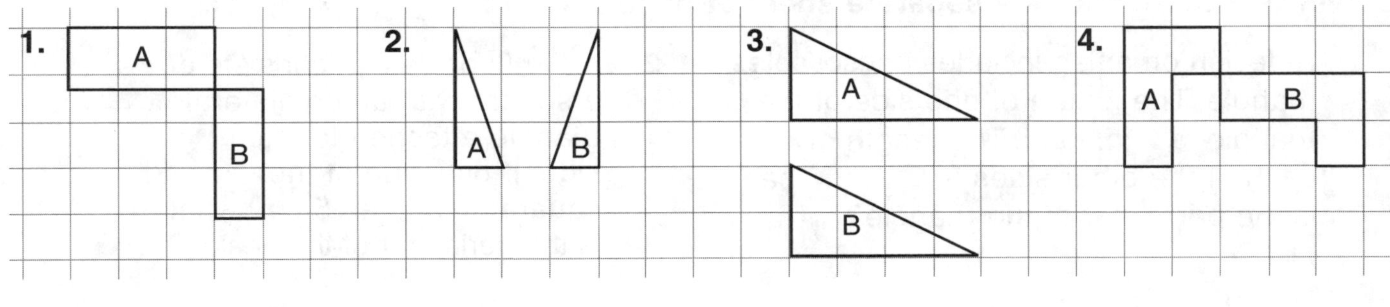

____________ ____________ ____________ ____________

Draw a second figure to show each movement.

5. Slide triangle *ABC* right 6 units and down 1 unit.

6. Flip triangle *DEF* across line.

7. Turn triangle *PQR* around vertex *R*.

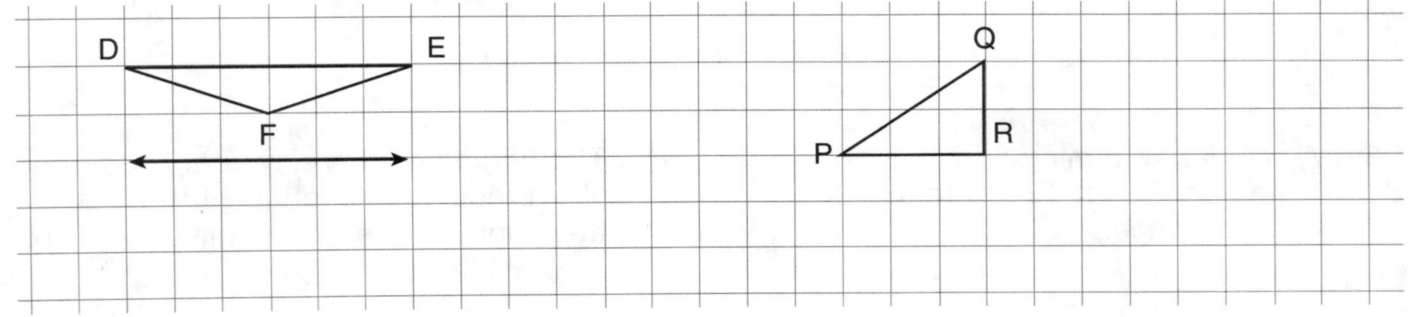

 Use with Lesson 8-12, text pages 290–291.

Problem-Solving Strategy: Use Formulas

Name ______________________________

Date ______________________________

The dimensions of a box are shown in the diagram. Find the perimeter of the top rectangular face.

60 cm
150 cm
90 cm

Write a formula for the perimeter of a rectangle.

$P = \ell + w + \ell + w$
$P = (2 \times \ell) + (2 \times w)$
$P = (2 \times 150 \text{ cm}) + (2 \times 90 \text{ cm})$
$P = 300 \text{ cm} + 180 \text{ cm}$
$P = 480 \text{ cm}$

Check:
$P = \ell + w + \ell + w$
$480 \text{ cm} \stackrel{?}{=} 150 \text{ cm} + 90 \text{ cm} + 150 \text{ cm} + 90 \text{ cm}$
$480 \text{ cm} = 480 \text{ cm}$

The perimeter of the top rectangular face is 480 cm.

Solve. Do your work on a separate sheet of paper.

1. A design on a flag includes an isosceles triangle. The length of one side of the triangle is 2 in. long. The length of each of the other sides is $2\frac{1}{4}$ in. What is the perimeter of the triangle?

____

2. A tile floor design consists of a square with an equilateral triangle attached to each side. If one side of the square measures 5 cm, what is the perimeter of the design?

____

3. A garden has the shape of a regular hexagon. If the perimeter of the garden is 138 ft, what is the length of one side of the garden?

____

4. Two of the angle measures for a triangle are 54° and 35°. What is the measure of the third angle?

____

5. Parallelogram *PQRS* has two pairs of congruent sides. The perimeter is 190 cm. If the length of $\overline{PQ}$ is 20 cm, what is the length of $\overline{QR}$?

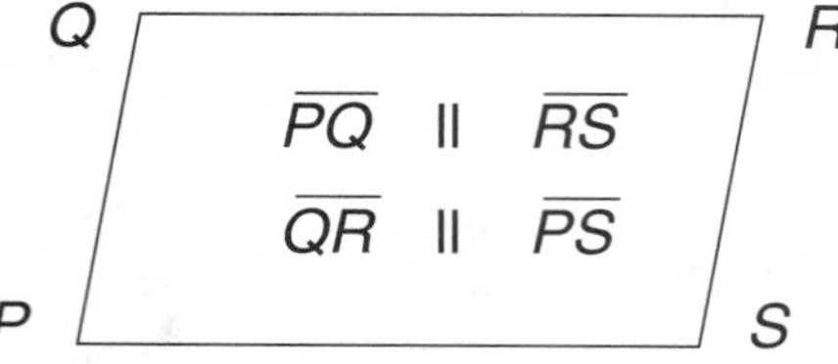

____

6. A circular swimming pool has a diameter of 24 ft. Estimate the circumference.

____

7. The perimeter of a play yard shaped like a regular octagon is 24 ft. What is the length of one side of the play yard?

____

 Use with Lesson 8-14, text pages 294–295.

Relating Customary Units of Length

Name ______________________

Date ______________________

3 ft 8 in. = ? in.

Think: 12 in. = 1 ft

3 ft = (3 × 12) in. = 36 in.

36 in. + 8 in. = 44 in.

So 3 ft 8 in. = 44 in.

138 in. = ? yd ___ in.

Think: 36 in. = 1 yd

(138 ÷ 36) yd = $3\frac{20}{36}$ yd

$3\frac{20}{36}$ yd = 3 yd 20 in.

So 138 in. = 3 yd 20 in.

List two classroom items that are:

1. longer than 1 inch but shorter than 1 foot.

   ____

2. longer than 1 foot but shorter than 1 yard.

   ____

3. longer than 1 yard.

   ____

Complete.

4. 7 ft 2 in. = ______ in.
5. 5 ft 5 in. = ______ in.
6. 8 yd 2 ft = ______ ft
7. 253 in. = ______ yd ______ in.
8. 137 ft = ______ yd ______ ft
9. 63 in. = ______ ft ______ in.
10. 99 in. = ______ ft ______ in.
11. 6790 ft = ______ mi ______ ft
12. 29 ft = ______ yd ______ ft

Measure each to the nearest inch, $\frac{1}{2}$ inch, $\frac{1}{4}$ inch, and $\frac{1}{8}$ inch.

13. ______________________
14. ______________________
15. ______________________
16. ______________________

PROBLEM SOLVING

17. The height of a doorway is 7 ft 4 in. Randall says it is 88 in. high. Kyoko says it is 2 yd 1 ft 6 in. high. James says it is 2 yd 16 in. high. Can they all be correct? Explain your answer.

    ____

Relating Customary Units of Capacity and Weight

Name ______________________________

Date ______________________________

Multiply to rename larger units as smaller units.
Divide to rename smaller units as larger units.

15 pt = ? qt
15 pt = (15 ÷ 2) qt
So 15 pt = $7\frac{1}{2}$ qt.

Think:
2 pt = 1 qt

3 lb 2 oz = ? oz
3 lb = (3 × 16) oz = 48 oz
48 oz + 2 oz = 50 oz
So 3 lb 2 oz = 50 oz.

Think:
1 lb = 16 oz

Complete. Do your work on a separate sheet of paper.

1. 4 gal = ______ qt	**2.** 19 qt = ____ gal ____ qt	**3.** 2 gal = ______ pt
4. 10 pt = ______ qt	**5.** 2 qt 1 pt = ______ c	**6.** $3\frac{1}{2}$ c = ______ fl oz
7. 27 c = ____ pt ____ c	**8.** 3 gal 1 qt = ______ pt	**9.** 18 fl oz = ____ c
10. 4 qt 1 c = ______ c	**11.** 1 gal 1 qt = ______ qt	**12.** 12 fl oz = ____ c
13. 5 T = _____ lb	**14.** 24 lb = ____ oz	**15.** 8000 oz = ____ T
16. 260 oz = ____ lb ____ oz	**17.** 5000 lb = ____ T ____ lb	**18.** 3 lb 9 oz = ____ oz
19. 4 T 625 lb = ____ lb	**20.** 12 lb 5 oz = ____ oz	**21.** 638 oz = ____ lb
22. 5 T 1908 lb = _____ lb	**23.** 490 oz = ____ lb ____ oz	**24.** 7300 lb = ____ T

Compare. Write <, =, or >.

25. 30 fl oz ____ 3 c 5 fl oz	**26.** 3 qt ____ 5 pt	**27.** 3 gal 2 qt ____ 16 qt
28. 10 qt ____ 2 gal 3 qt	**29.** 32 pt ____ 15 qt 1 pt	**30.** 2 c 7 fl oz ____ 21 fl oz
31. 6000 lb ____ 3 T	**32.** 3 lb ____ 45 oz	**33.** 4010 lb ____ 2 T 1 lb
34. 11 lb 5 oz ____ 180 oz	**35.** 5 lb 17 oz ____ 4 lb 33 oz	**36.** 111 oz ____ 7 lb 9 oz

PROBLEM SOLVING

37. A punch recipe calls for $2\frac{1}{2}$ qt orange juice. William needs to make 5 times as much punch for a banquet. Will he have enough orange juice if he buys 4 gallons?

____

38. Darla is shipping a statue that weighs 120 oz. Can she safely ship the statue in a box that can hold a maximum of 8 pounds?

____

Temperature

Name ______________________

Date ______________________

	Water boils	Water freezes
Fahrenheit:	212°F	32°F
Celsius:	100°C	0°C

Starting Temperature: 20°C
Change: falls 30°C
Final Temperature: ⁻10°C

Circle the most reasonable temperature for each.

		a.	b.	c.
1.	going swimming	**a.** 32°F	**b.** 60°F	**c.** 92°F
2.	building a snowman	**a.** ⁻60°F	**b.** 10°F	**c.** 50°F
3.	roasting a chicken	**a.** 15°C	**b.** 150°C	**c.** 32°C
4.	body temperature	**a.** 37°C	**b.** 10°C	**c.** 70°C

Write the final temperature.

5. Celsius

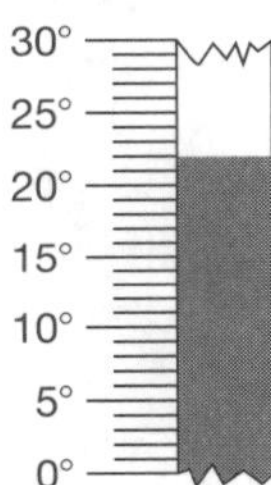

Change: falls 22°
Final Temperature: ______

6. Fahrenheit

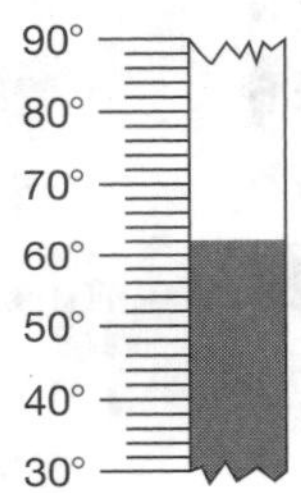

Change: rises 8°
Final Temperature: ______

7. Fahrenheit

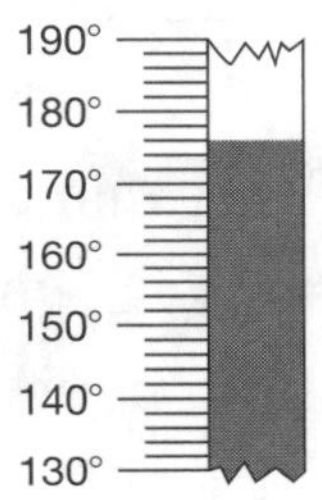

Change: rises 14°
Final Temperature: ______

8. Celsius

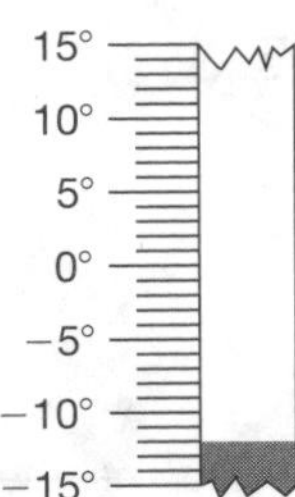

Change: rises 21°
Final temperature: ______

9. Fahrenheit

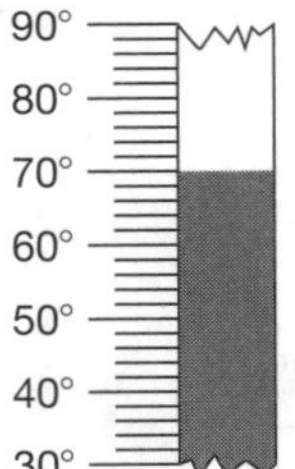

Change: falls 24°
Final temperature: ______

10. Celsius

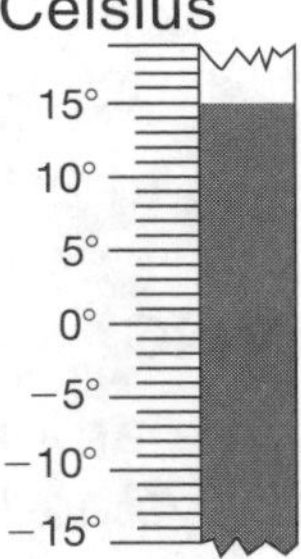

Change: falls 20°
Final temperature: ______

Complete.

11.

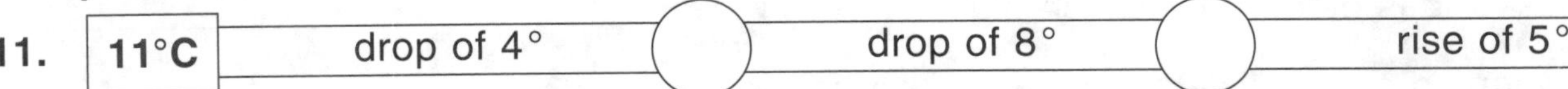

12.

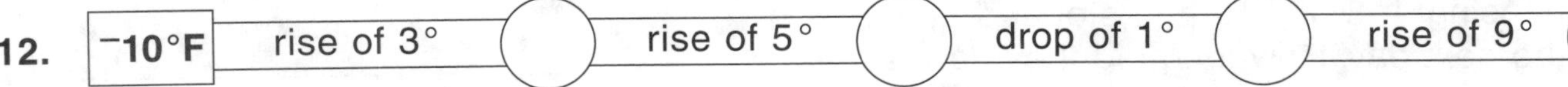

Use with Lesson 9-4, text pages 308–309.

Units of Time

Name ______________________

Date ______________________

60 s = 1 min
60 min = 1 h
24 h = 1 d
7 d = 1 wk
12 mo = 1 y
365 d = 1 y
100 y = 1 cent.

Multiply to rename larger units as smaller units.

Divide to rename smaller units as larger units

Complete.

1. 48 mo = ______ y
2. 10 y = ______ mo
3. 4 h = ______ min
4. 56 d = ______ wk
5. 240 s = ______ min
6. 168 h = ______ d
7. 120 min = ______ h
8. 36 mo = ______ y
9. 2 wk = ______ d
10. 480 s = ______ min
11. 12 d = ______ h
12. 3 h = ______ min
13. 153 min = ______ h ______ min
14. 18 mo = ______ y ______ mo
15. 2 wk 4 d = ______ d
16. 1 h 23 min = ______ min
17. 115 s = ______ min ______ s
18. 130 d = ______ wk ______ d
19. 2 y 3 mo = ______ mo
20. 4 min 32 s = ______ s
21. 630 y = ______ cent. ______ y
22. 45 h = ______ d ______ h

Find the elapsed time.

23. from 1:30 P.M. to 11:45 P.M. ______________
24. from 8:20 A.M. to 1:15 P.M. ______________
25. from 10:30 P.M. to 12:10 A.M. ______________
26. from 6:24 A.M. to 2:13 P.M. ______________

PROBLEM SOLVING

27. One flight time around the world was 63 hours. How many days and hours was this? ______________
28. One boat arrived at 1:33 P.M. and a second boat arrived at 3:43 P.M. How much time elapsed between the two arrivals? ______________
29. Susan is 1 y 4 mo old. William is 18 mo old. Which child is older? how much older? ______________
30. The winning couple in a dance marathon began dancing at 8:35 P.M. and stopped at 4:12 A.M. the next day. How long did they dance? ______________

 Use with Lesson 9-5, text pages 310–311.

Time Zones

Name ______________________

Date ______________________

West ➤- ➤ East

Pacific	**Mountain**	**Central**	**Eastern**
7:56 A.M.	8:56 A.M.	9:56 A.M.	10:56 A.M.

As you travel from west to east, the time in each zone is one hour later.

Complete.

	Time Zone			
	Pacific	**Mountain**	**Central**	**Eastern**
1.				8:00 P.M.
2.		11:30 A.M.		
3.			2:43 P.M.	
4.	7:27 A.M.			

	City	Time Zone			
		Pacific	**Mountain**	**Central**	**Eastern**
5.	Portland, OR	6:29 A.M			
6.	Portland, ME				7:14 P.M.
7.	Denver, CO		12:00 noon		
8.	Dallas, TX			12:00 midnight	

PROBLEM SOLVING

9. Thomas flies from Boston to Seattle in 5 hours. If he leaves at 9:00 A.M. Eastern time, what time will it be in Pacific time when he arrives in Seattle? ______________________

10. A company schedules a conference call at 11:00 A.M. Central time. At what time does the call take place in Eastern time? in Mountain time? in Pacific time? ______________________

11. James lives in Los Angeles. He calls his sister at 10:30 A.M. Pacific time. His sister lives in the Central time zone. At what time does she receive the call? ______________________

12. Alicia lives in Portland, ME. She calls her mother in the Mountain time zone at 9:00 P.M. Eastern time. At what time does her mother receive the call? ______________________

Computing Customary Units

Name ______________________

Date ______________________

Add.

$$\begin{array}{r} 2 \text{ yd } 2 \text{ ft} \\ + \ 3 \text{ yd } 2 \text{ ft} \\ \hline 5 \text{ yd } 4 \text{ ft} \end{array} = 6 \text{ yd } 1 \text{ ft}$$

4 ft = **1 yd 1 ft**

Subtract.

$$\begin{array}{r} 5 \text{ gal } 2 \text{ qt} \\ - \ 1 \text{ gal } 3 \text{ qt} \\ \hline \end{array}$$

= **4** gal + **1** gal 2 qt

= **4** gal **6** qt

$$\begin{array}{r} 4 \text{ gal } 6 \text{ qt} \\ - \ 1 \text{ gal } 3 \text{ qt} \\ \hline 3 \text{ gal } 3 \text{ qt} \end{array}$$

Add.

1. 9 lb 3 oz + 2 lb 7 oz

2. 7 yd 2 ft + 3 yd 1 ft

3. 8 qt 1 pt + 1 qt 1 pt

4. 2 gal 2 qt + 4 gal 2 qt

5. 6 wk 4 d + 2 wk 9 d

6. 21 ft 9 in. + 9 ft 8 in.

Subtract.

7. 8 d 7 h − 5 d 3 h

8. 11 gal 3 qt − 11 gal 1 qt

9. 3 wk 5 d − 2 wk 5 d

10. 5 yd 12 in. − 2 yd 22 in.

11. 4 qt − 2 qt 1 pt

12. 6 T 240 lb − 2 T 680 lb

Find the sum or difference.

13. 5 pt 1 c + 3 pt 1 c = ______________

14. 5 d 8 h − 4 d 20 h = ______________

15. 3 h 39 min + 4 h 57 min = ________

16. 9 h 16 min − 3 h 38 min = ________

17. 4 ft 11 in. + 6 ft 8 in. = __________

18. 6 lb 2 oz − 1 lb 7 oz = __________

PROBLEM SOLVING

19. Elena has a piece of lumber 7 yd long. How much must she cut off to have 2 yd 1 ft of lumber left? ______________

20. Sam watched a music video on television that lasted 17 minutes and a movie that lasted 1 hour and 39 minutes. How long did Sam watch television? ______________

 Use with Lesson 9-7, text pages 314–315.

Problem-Solving Strategy: Multi-Step Problem

Name ____________________

Date ____________________

Lucas made 2 bottles of grape punch and 3 bottles of orange punch for a picnic. Each bottle of grape punch held 1 qt 1 pt, and each bottle of orange punch held 2 qt 1 pt. How much punch did he make in all?

Add to find the amount of each punch.

```
  1 qt 1 pt                      2 qt 1 pt
+ 1 qt 1 pt                      2 qt 1 pt
  2 qt 2 pt grape punch        + 2 qt 1 pt
                                 6 qt 3 pt of orange punch
```

Change the units: 2 pt = 1 qt.

3 qt grape punch — 7 qt 1 pt orange punch

Add to find the total amount of punch

```
  3 qt
+ 7 qt 1 pt
 10 qt 1 pt
```

Lucas made 10 qt 1 pt of punch.

Solve. Do your work on a separate sheet of paper.

1. Lori used 2 ft 4 in. of ribbon for each dress she made and 1 ft 8 in. for each skirt. She made 3 dresses and 4 skirts. How much ribbon did she use in all?

   ____

2. There are 4 dozen blueberry muffins in the display case at Betty's Bakery. There are also 7 trays of bran muffins, with 8 muffins in each tray. How many muffins are there altogether?

   ____

3. Peter rented a movie that has a running time of 1 hour 55 minutes. He began watching the movie at 6:30 P.M. He turned it off for 15 minutes and then continued watching the movie to the end. What time did he finish watching the movie?

   ____

4. Jacque recorded the noon temperatures every day last week and found the average temperature to be 20°C. This week he recorded the following temperatures: 18°C, 16°C, 13°C, 15°C, 14°C, 16°C, and 13°C. How many degrees did the average temperature decrease?

   ____

5. Eleanor bought 5 bags of peanuts and 3 bags of raisins. Each bag of peanuts weighed 1 lb 2 oz, and each bag of raisins weighed 2 lb 9oz. What was the total weight of the peanuts and raisins Eleanor bought?

   ____

Decimal Sense

Name ______________________

Date ______________________

As with whole numbers, a greater decimal is located to the right of a lesser decimal on a number line.

Point *A* represents 1.7

Point *B* represents 2.1

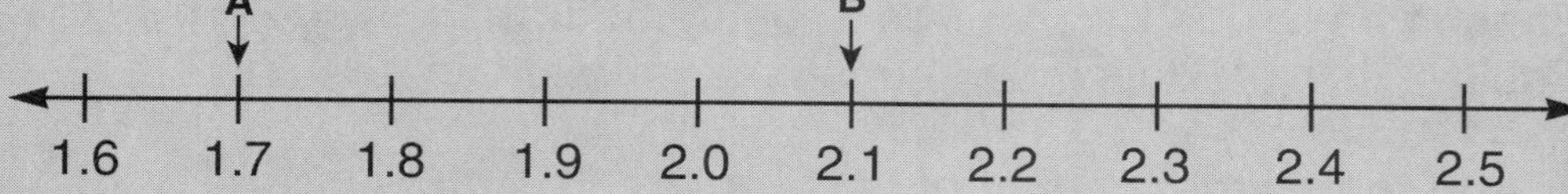

Point *A* represents 0.71

Point *B* represents 0.745

Point *C* represents 0.782

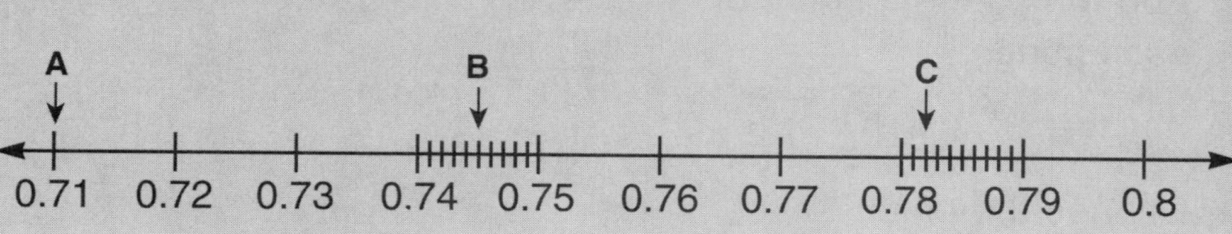

Name the decimal represented by each point on each number line.

1. A B C 0.5 0.6

A represents __________

B represents __________

C represents __________

2. A B C 0.56 0.57

A represents __________

B represents __________

C represents __________

3. A B C 8.8 8.9

A represents __________

B represents __________

C represents __________

4.

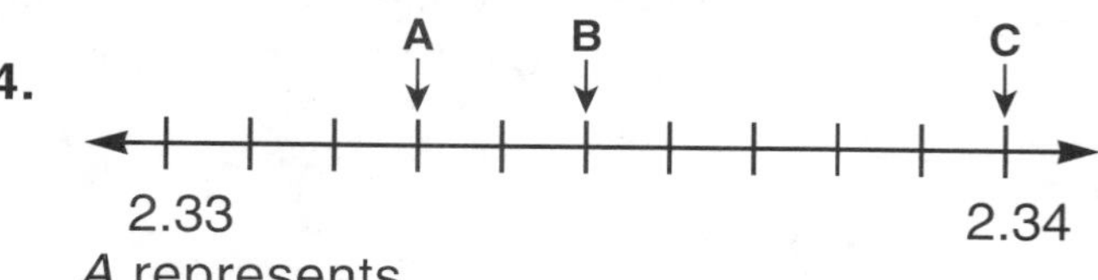

A represents __________

B represents __________

C represents __________

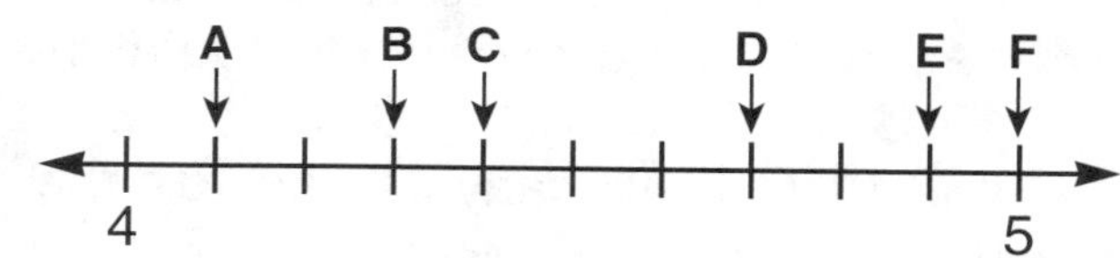

5. *A* __________ **6.** *D* __________ **7.** *B* __________ **8.** *E* __________

Name the point represented by each decimal.

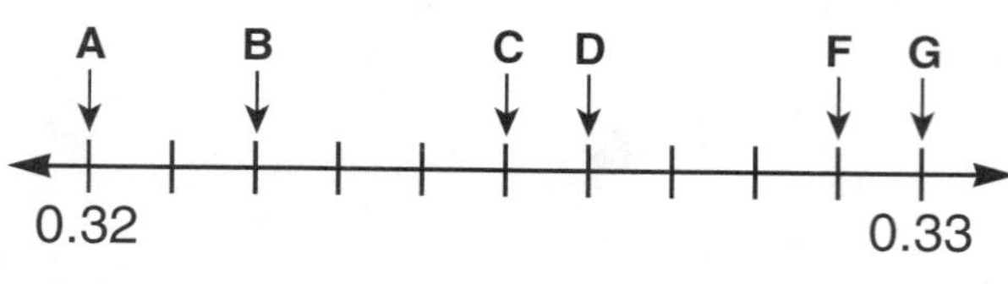

H I J K L M 4.41 4.42

9. 0.329 ______ **10.** 4.41 ______ **11.** 0.326 ______ **12.** 4.415 ______

13. 4.413 ______ **14.** 0.322 ______ **15.** 4.419 ______ **16.** 0.32 ______

 Use with Lesson 10-1, text pages 326–327.

Decimals and Place Value

Name ______________________

Date ______________________

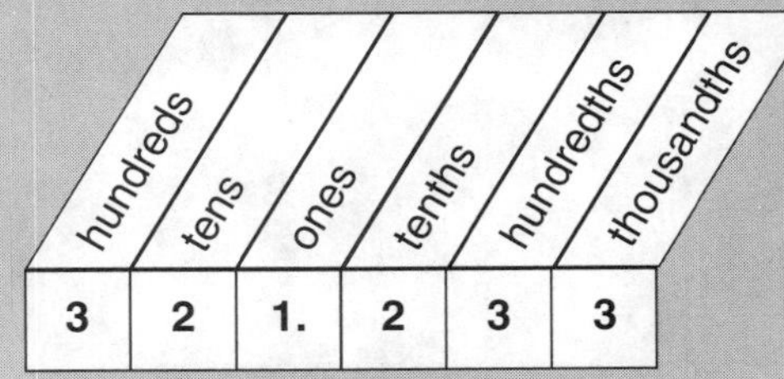

Standard Form:
321.233

Read:
Three hundred twenty-one *and* two hundred thirty-three thousandths

Expanded Form:
300 + 20 + 1+ 0.2+ 0.03 +

Write each number in standard form.

1. ten thousandths ________
2. six hundred two thousandths ________
3. three and two tenths ________
4. sixteen and forty-five hundredths ________

Write each number in expanded form.

5. 0.329 ______________________________
6. 2.03 ______________________________
7. 4.007 ______________________________
8. 23.45 ______________________________

Write each in standard form.

9. 8 + 0.2 + 0.02 + 0.004 ________
10. 4 + 0.5 + 0.003 ________
11. 0.8 + 0.03 + 0.001 ________
12. 42 + 0.09 ________
13. 2 + 0.1 + 0.05 + 0.005 ________
14. 16 + 0.7 + 0.07 + 0.001 ________

Circle the letter of the correct answer.

15. In the number 4.832, the 8 means:
    a. 8 × 1 **b.** 8 × 0.01 **c.** 8 × 0.1 **d.** 8 × 0.001
16. In the number 21.673, the 3 means:
    a. 2 × 10 **b.** 3 × 0.01 **c.** 3 × 0.100 **d.** 3 × 0.001
17. In the number 7.394, the 9 means:
    a. 9 × 0.01 **b.** 9 × 0.001 **c.** 9 × 1 **d.** 9 × 0.100
18. In the number 79.43, the 9 means:
    a. 9 × 0.100 **b.** 9 × 1 **c.** 9 × 0.01 **d.** 9 × 1000
19. In the number 5.064, the 6 means:
    a. 6 × 1 **b.** 6 × 0.01 **c.** 6 × 0.001 **d.** 6 × 0.1

Use with Lesson 10-2, text pages 328–329.

Adding Decimals

Name ______________________________

Date ______________________________

Add: 0.49 + 0.7 + 0.28 = ?

To add decimals:
- Line up the decimal points.
- Add as with whole numbers.
- Write the decimal point in the sum.

```
   1 1
   0.49
   0.70  <---  0.7 = 0.70
 + 0.28
 ------
   1.47
```

Add.

1. 0.04 + 0.23
2. 0.73 + 0.25
3. 0.5 + 0.5
4. 0.72 + 0.2
5. 0.7 + 0.4
6. 0.38 + 0.21
7. 0.5 + 0.68
8. 0.35 + 0.7
9. 0.47 + 0.09
10. 0.13 + 0.08

Find the sum.

11. 0.3 + 0.68 + 0.97
12. 0.63 + 0.04 + 0.15
13. 0.7 + 0.63 + 0.08
14. 0.21 + 0.14 + 0.07
15. 0.24 + 0.2 + 0.91

Align and add.

16. 0.04 + 0.64 + 0.13 __________
17. 0.27 + 0.56 + 0.38 __________
18. 0.74 + 0.19 + 0.72 __________
19. 0.47 + 0.6 + 0.09 __________
20. 0.08 + 0.05 + 0.11 __________
21. 0.3 + 0.09 + 0.86 __________

Compare. Write <, =, or >.

22. 0.04 + 0.74 + 0.13 ______ 0.08 + 0.64 + 0.32
23. 0.27 + 0.31 + 0.72 ______ 0.38 + 0.03 + 0.7
24. 0.471 + 0.563 + 0.32 ______ 0.573 + 0.581 + 0.2

PROBLEM SOLVING

25. A triangle has 3 equal sides. One side has a length of 0.85 cm. What is the perimeter of the triangle?

____

26. A quadrilateral has 4 equal sides. Two of the sides have a total length of 0.90 m. What is the perimeter?

____

Estimate Decimal Sums

Name ______________________

Date ______________________

Estimate: 2.9 + 3.06

Front-End

$$\begin{array}{r} 2.9 \\ +\ 3.06 \\ \hline \end{array}$$

about 5.00

Rounding

$$\begin{array}{rcr} 2.9 & \longrightarrow & 3 \\ +\ 3.06 & \longrightarrow & +\ 3 \\ \hline & & \text{about } 6 \end{array}$$

So the exact sum is between 5 and 6.

Estimate the sum by rounding.

1. $\begin{array}{r} 2.71 \\ 4.12 \\ +\ 6.5 \\ \hline \end{array}$

2. $\begin{array}{r} 6.3 \\ 5.92 \\ +\ 2.37 \\ \hline \end{array}$

3. $\begin{array}{r} 6.5 \\ 1.624 \\ +\ 7.328 \\ \hline \end{array}$

4. $\begin{array}{r} 0.73 \\ 0.9 \\ +\ 3.321 \\ \hline \end{array}$

Estimate the sum. Use front-end estimation.

5. $\begin{array}{r} 0.35 \\ 0.21 \\ +\ 0.7 \\ \hline \end{array}$

6. $\begin{array}{r} 5.45 \\ 2.13 \\ +\ 2.97 \\ \hline \end{array}$

7. $\begin{array}{r} 5.34 \\ 1.358 \\ +\ 2.824 \\ \hline \end{array}$

8. $\begin{array}{r} 4.296 \\ 1.683 \\ +\ 9.198 \\ \hline \end{array}$

Estimate by both rounding and front-end estimation.
Between what two numbers will the exact sum be?

9. $\begin{array}{r} 5.8 \\ +\ 2.63 \\ \hline \end{array}$

____

____

10. $\begin{array}{r} 7.82 \\ +\ 2.9 \\ \hline \end{array}$

____

____

11. $\begin{array}{r} 42.82 \\ +\ 25.163 \\ \hline \end{array}$

____

____

12. $\begin{array}{r} 5.66 \\ 3.628 \\ +\ 6.089 \\ \hline \end{array}$

____

____

Estimate the sum.

13. $\begin{array}{r} 0.38 \\ +\ 0.7 \\ \hline \end{array}$

14. $\begin{array}{r} 0.19 \\ +\ 0.136 \\ \hline \end{array}$

15. $\begin{array}{r} 0.471 \\ +\ 0.563 \\ \hline \end{array}$

16. $\begin{array}{r} 2.369 \\ +\ 1.247 \\ \hline \end{array}$

17. $\begin{array}{r} 12.902 \\ +\ 18.24 \\ \hline \end{array}$

18. $\begin{array}{r} 36.134 \\ 12.1 \\ +\ 63.08 \\ \hline \end{array}$

19. $\begin{array}{r} 24.491 \\ 16.032 \\ +\ 13.92 \\ \hline \end{array}$

20. $\begin{array}{r} 1.56 \\ 8.002 \\ +\ 1.1 \\ \hline \end{array}$

21. $\begin{array}{r} 10.09 \\ 29.9 \\ +\ 30.321 \\ \hline \end{array}$

22. $\begin{array}{r} 3.62 \\ 3.090 \\ +\ 1.01 \\ \hline \end{array}$

More Adding Decimals

Name ______________________

Date ______________________

Add: 5.72 + 0.009 + 27.8 = ?

- Line up the decimal points.
- Add as usual.
- Write the decimal point in the sum.

```
   1 1
   5.720   <--  5.72 = 5.720
   0.009
+ 27.800   <--  27.8 = 27.800
  ------
  33.529
```

Estimate. Then find the sum.

1. 4.87 + 6.9
2. 12 + 8.34
3. 9.7 + 9.75
4. 0.38 + 0.8
5. 5.09 + 4.93
6. 23.1 + 4.042
7. 7.395 + 8.46
8. 36.05 + 6.2
9. 58.912 + 0.066
10. 2.736 + 0.5
11. 1.17 + 8.064
12. 17.4 + 3.909
13. 0.627 + 8.58
14. 5.298, 0.63, + 4.04
15. 20.6, 1.379, + 24.42
16. 0.67, 9.209, + 1.8
17. 3.841, 0.933, + 5.5

Align and add.

18. 4.3 + 62.04 + 8.7 = ________
19. 0.008 + 13.6 + 1.775 = ________
20. 1.838 + 2.5 + 3.32 = ________
21. 64.9 + 0.05 + 3.833 = ________
22. 44 + 0.736 + 8.1 = ________
23. 26.05 + 2.514 + 8.8 = ________

PROBLEM SOLVING

24. Jenelle scored 8.913, 8.099, and 9.2 in three gymnastics events. What was her total score? ________________

25. In a charity relay walk, Mike walks 2.7 km and hands his baton to Kristen. Kristen walks 2.45 km and hands her baton to Tommy, who walks 1.98 km. How much distance do they cover? ________________

 Use with Lesson 10-5, text pages 334–335.

Subtracting Decimals

Name ____________________

Date ____________________

Subtract: 0.9 − 0.85 = ?

To subtract decimals:

- Line up the decimal points.
- Subtract as with whole numbers.
- Write the decimal point

```
   8 10
 0.9 0   ←  0.9 = 0.90
−0.8 5
 -----
 0.0 5
```

Find the difference.

1. 0.6 − 0.4	**2.** 0.8 − 0.3	**3.** 0.35 − 0.26	**4.** 0.92 − 0.09	**5.** 0.77 − 0.54
6. 0.99 − 0.7	**7.** 0.58 − 0.4	**8.** 0.72 − 0.1	**9.** 0.33 − 0.2	**10.** 0.88 − 0.6
11. 0.9 − 0.65	**12.** 0.7 − 0.53	**13.** 0.8 − 0.37	**14.** 0.5 − 0.29	**15.** 0.6 − 0.18

Align and subtract.

16. 0.54 − 0.3 = ________

17. 0.8 − 0.46 = ________

18. 0.75 − 0.44 = ________

19. 0.7 − 0.5 = ________

20. 0.4 − 0.39 = ________

21. 0.5 − 0.26 = ________

22. 0.69 − 0.08 = ________

23. 0.07 − 0.03 = ________

PROBLEM SOLVING

24. What is the difference between 0.38 and 0.09? ____________________

25. How much less than 0.7 is 0.25? ____________________

26. Rafael lives 0.6 km from school. Marisa lives 0.67 km from school. Who lives farther from school? how much farther? ____________________

27. Feng walks 0.8 km to school. Valerie walks 0.75 km to school. How much farther does Feng have to walk? ____________________

 Use with Lesson 10-6, text pages 336–337.

Estimate Decimal Differences

Name ______________________

Date ______________________

Estimate: 4.75 − 2.36

Front-End	Rounding
4.75 − 2.36 about 2.00	4.75 → 5 − 2.36 → − 2 about 3

So the exact difference is between 2 and 3.

Estimate the difference by rounding.

1. 0.553 − 0.274
2. 0.5 − 0.385
3. 0.58 − 0.316
4. 4.13 − 2.98

Estimate the difference. Use front-end estimation.

5. 0.85 − 0.24
6. 0.517 − 0.284
7. 5.46 − 3.65
8. 30.34 − 12.06

Estimate by both rounding and front-end estimation. Between what two numbers will the exact difference be?

9. 8.25 − 3.76
10. 6.34 − 3.76
11. 69.372 − 12.53
12. 45.326 − 23.917

Estimate.

13. 0.29 − 0.14
14. 0.57 − 0.32
15. 0.34 − 0.19
16. 42.09 − 28.245
17. 0.538 − 0.295
18. 6.89 − 1.148
19. 52.385 − 27.06
20. 33.79 − 18.82
21. 0.89 − 0.6
22. 0.251 − 0.075

PROBLEM SOLVING

23. Susan used 1.75 m of a 3.5-meter length of yarn. She needs 1 m of the yarn for a project. Does she have enough yarn left? ______________________

 Use with Lesson 10-7, text pages 338–339.

More Subtracting Decimals

Name ______________________

Date ______________________

Subtract: 5.2 − 4.738 = ?

- Line up the decimal points.
- Subtract as usual.
- Write the decimal point in the difference.

```
     11 9
  4 10 10 10
  5 . 2 0 0   ←  5.2 = 5.200
− 4 . 7 3 8
  0 . 4 6 2
```

Estimate. Then find the difference.

1. 8.7 − 4.3
2. 3.05 − 0.17
3. 6.26 − 3.91
4. 4.097 − 1.625
5. 9.342 − 5.806
6. 7.52 − 5.8
7. 12.25 − 8.1
8. 6.198 − 4.28
9. 9.077 − 8.35
10. 3.281 − 1.9
11. 5.5 − 2.37
12. 10.91 − 6.358
13. 8.2 − 4.019
14. 7.7 − 5.05
15. 6.04 − 3.549
16. 13.2 − 11.073
17. 54.1 − 0.84
18. 22.16 − 9.504
19. 33.5 − 3.55
20. 17 − 15.606

Align and subtract.

21. 45.6 − 3.67 = ______
22. 101.75 − 47.8 = ______
23. 5.51 − 4.936 = ______
24. 55.7 − 18.586 = ______
25. 86.13 − 0.999 = ______
26. 99.9 − 9.123 = ______

PROBLEM SOLVING

27. The distance from the library to the school is 2.7 km. The distance from the library to Town Hall is 2.086 km. Which is the longer distance? by how much? ______

28. Malik has painted 3.75 m of a fence that is 10.5 m long. How much of the fence is left for him to paint? ______

Problem-Solving Strategy: Extra Information

Name ______________________

Date ______________________

It takes Maya 20 minutes to walk to the school and the same amount of time to walk home. It is 0.8 mile from her home to her school. How much time does she spend walking to and from school in 5 days?

There is more information than you need.

First, find how much time she spends walking per day. Then, multiply the answer to find how much time she spends walking in five days.

$2 \times 20 = 40$
$5 \times 40 = 200$

Maya spends 200 min, or 3 h 20 min walking to and from school in 5 days.

Solve. Do your work on a separate sheet of paper.

1. Yuko is 1.35 m tall. His mother is 0.25 m taller than Yuko, and his father is 1.8 m tall. How tall is Yuko's mother?

   ____

2. Pamela buys two 60-ounce cartons of pineapple juice at $1.99 each and two $\frac{1}{2}$-gallon cartons of orange juice at $1.99 each. How much does she pay for the juice?

   ____

3. Vinnie and Ellen go to different schools. There are 38 students in the 4th grade of Vinnie's school, 42 in the 5th grade, and 39 in the 6th grade. There are 127 students in grades 4, 5, and 6 of Ellen's school. How many students are in grades 4, 5, and 6 of Vinnie's school?

   ____

4. Mr. and Mrs. Ortiz rode their bicycles. It took them 8 minutes to ride 1.2 miles to the park, 6 minutes to ride 0.8 miles to the post office, and 7 minutes to ride 1 mile back home. How many miles did they ride?

   ____

5. The Science Club noted that the temperature rose from 33° F at noon on Monday to 41° F at 2:30 P.M. On Tuesday the temperature rose from 30° F at noon to 39° F at 2:30 P.M. What was the increase in temperature on Monday?

   ____

6. Perry buys 2 cans of tomato soup at $.57 each, 1 box of crackers for $2.49, and 3 cans of extra-chunky vegetable soup at $1.29 each. How much does he pay for the soup?

   ____

 Use with Lesson 10-10, text pages 344–345.

Multiply by 10, 100, and 1000

Name ____________________

Date ____________________

Multiply:

10 × 0.437 = ?	**10** × 0.437 = 4.37	← 1 zero. Move 1 place to the right.
100 × 0.058 = ?	**100** × 0.058 = 5.8	← 2 zeros. Move 2 places to the right.
1000 × 0.007 = ?	**1000** × 0.007 = 7	← 3 zeros. Move 3 places to the right.
1000 × 9.6 = ?	**1000** × 9.600 = 9600	← 3 zeros. Move 3 places to the right. Write 2 zeros as placeholders.

Write the product.

1. 10 × 0.637 = ______
100 × 0.637 = ______
1000 × 0.637 = ______

2. 10 × 0.003 = ______
100 × 0.003 = ______
1000 × 0.003 = ______

3. 10 × 1.008 = ______
100 × 1.008 = ______
1000 × 1.008 = ______

4. 10 × 0.12 = ______
100 × 0.12 = ______
1000 × 0.12 = ______

5. 10 × 6.47 = ______
100 × 6.47 = ______
1000 × 6.47 = ______

6. 10 × 0.9 = ______
100 × 0.9 = ______
1000 × 0.9 = ______

Multiply.

7. 10 × 0.03 = _____
8. 100 × 0.237 = _____
9. 1000 × 0.036 = _____
10. 100 × 0.004 = _____
11. 10 × 7.5 = _____
12. 100 × 24.8 = _____
13. 1000 × 0.001 = _____
14. 10 × 6.841 = _____
15. 1000 × 4.3 = _____

Find the missing factor.

16. _____ × 0.703 = 70.3
17. _____ × 0.084 = 0.84
18. _____ × 0.523 = 523
19. 10 × _____ = 15.6
20. 100 × _____ = 780
21. 1000 × _____ = 2007

Find the product.

22. 20 × 0.7 = ___
23. 50 × 0.3 = ___
24. 400 × 0.4 = ___
25. 700 × 0.7 = ___
26. 600 × 0.2 = ___
27. 300 × 0.8 = ___
28. 90 × 0.6 = ___
29. 60 × 0.1 = ___

PROBLEM SOLVING

30. Grace bought a 10 lb turkey at $1.19 per pound. How much did she spend? ____________

31. If Debbie walked 1.026 mi each day for 100 days, how far did she walk? ____________

Estimating Decimal Products

Name ______________________

Date ______________________

Estimate: 4.2 × 0.843 = ?	Estimate: 56.93 × 0.452 = ?	Estimate: 1.03 + 1.2 + 0.9 = ?
0.843 → 0.8 × 4.2 → × 4 about 3.2	0.452 → 0.5 × 56.93 → × 60 about 30.0	1.03 + 1.2 + 0.9 ↓ ↓ ↓ 1 + 1 + 1 3 × 1 = 3
Both factors are rounded down. The actual product is greater than 3.2.	Both factors are rounded up. The actual product is less than 30.	So the sum is about 3.

Estimate each product. Then tell whether the actual product is *greater than* or *less than* the estimated product.

1. 5.21 × 4.3

2. 0.98 × 7.61

3. 4.52 × 3.8

4. 1.208 × 5.1

5. 2.534 × 8.63

6. 9.127 × 6.04

7. 4.008 × 2.09

8. 5.375 × 9.25

9. 18.54 × 0.78

10. 61.203 × 0.42

11. 72.842 × 0.63

12. 38.002 × 0.49

Estimate the sum. Use clustering.

13. 6.13 + 5.9 + 6.008 ______

14. 0.82 + 0.8 + 0.84 + 0.79 + 0.78 ______

15. 42.1 + 39.5 + 41.4 + 40.75 + 39.631 ______

16. 7.004 + 6.75 + 6.891 + 7.2 + 7.5 + 7.4 ______

17. \$.25 + \$.32 + \$.34 + \$.28 ______

18. \$.07 + \$.13 + \$.06 + \$.13 + \$.14 ______

19. \$1.05 + \$.98 + \$.95 + \$1.02 ______

20. \$.45 + \$.51 + \$.54 + \$.49 + \$.48 ______

PROBLEM SOLVING

21. Pilar packs 6.1 cartons per hour. At this rate, about how many cartons can she pack in 2.25 hours? ______________

22. Mark prints 17.5 pages on his printer each hour. About how many pages can he print in 7.8 hours? ______________

23. Jane weighs samples of a salt and water solution. The samples weigh 3.5 g, 4.01 g, 4.3 g, 3.75 g, and 3.9 g. About how much do all the samples weigh together? ______________

 Use with Lesson 11-2, text pages 354–355.

Multiplying Decimals by Whole Numbers

Name ________________

Date ________________

7 × 0.5 = ?

$$\begin{array}{r} 0.5 \\ \times\ 7 \\ \hline 3.5 \end{array}$$

1 decimal place

23 × 3.12 = ?

$$\begin{array}{r} 3.12 \\ \times\ 23 \\ \hline 936 \\ 6240 \\ \hline 71.76 \end{array}$$

2 decimal places

31 × $15.25 = ?

$$\begin{array}{r} \$15.25 \\ \times\ 31 \\ \hline 1525 \\ 45750 \\ \hline \$472.75 \end{array}$$

2 decimal places with money

Write the decimal point in each product.

1. 3.7 × 2 = 74
2. 8.23 × 15 = 12345
3. 0.31 × 7 = 217
4. 6.54 × 12 = 7848
5. 3.264 × 63 = 205632

Estimate. Then find the product.

6. 0.5 × 19
7. 0.3 × 42
8. 0.76 × 28
9. 1.04 × 16
10. 0.089 × 47
11. 3.059 × 14
12. 3.427 × 53
13. $.34 × 12
14. $6.37 × 45
15. $23.82 × 68

Multiply.

16. 8 × 0.4 = ______
17. 2 × 0.315 = ______
18. 7 × 0.86 = ______
19. 5 × 0.236 = ______
20. 4 × 4.9 = ______
21. 3 × 1.064 = ______
22. 7 × 8.703 = ______
23. 12 × 12.8 = ______
24. 23 × 16.5 = ______
25. 48 × 22.731 = ______
26. 14 × $25.16 = ______
27. 87 × $1.93 = ______
28. two times eighteen hundredths = ______
29. six times thirty-four thousandths = ______

PROBLEM SOLVING

30. A can contains 6.5 oz of tuna fish. How many ounces do 15 cans contain? ______

31. Marilyn bought 6 cans of tuna at $1.29 a can. How much did she spend? ______

Multiplying Decimals by Decimals

Name ______________________

Date ______________________

0.5 × 0.7 = ?

35 out of 100 squares are marked off.

So 0.5 × 0.7 = 0.35.

Multiply: 4.2 × 3.08 = ?

```
   3.08  ← 2 decimal places
 ×  4.2  ← 1 decimal place
    616
  12320
 12.936  ← 3 decimal places
```

Use the diagram to complete each statement.

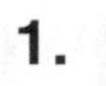

1. 

0.3 × 0.6 = ______

2. 

______ × ______ = 0.32

3. 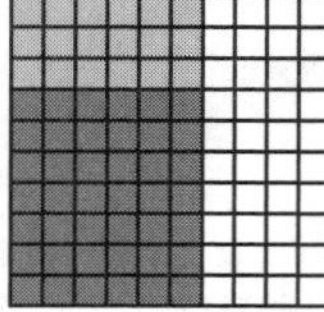

0.7 × ______ = 0.42

Write the decimal point in each product.

4.	5.	6.	7.	8.
5.9	3.76	7.13	15.2	18.2
× 0.6	× 3.4	× 0.4	× 0.03	× 0.05
3 54	12 784	2 852	0 456	0 910

Multiply.

9.	10.	11.	12.	13.
3.1	20.5	0.37	36.2	24.6
× 0.7	× 3.4	× 2.1	× 0.83	× 0.3

14.	15.	16.	17.	18.
13.3	3.62	2.07	6.13	3.87
× 0.16	× 5.4	× 3.6	× 0.8	× 48

PROBLEM SOLVING

19. A nature trail is 24.8 km long. Lizzie has hiked 0.65 of the length. How many kilometers has Lizzie hiked? ______

20. Jim paid $3.25 per pound for lamb chops. How much did he pay for a 2.2 lb package? ______

Zeros in the Product

Name ______________________

Date ______________________

$0.3 \times 0.02 = \underline{?}$

Multiply as with whole numbers.

$$\begin{array}{r} 0.02 \\ \times\ 0.3 \\ \hline 6 \end{array}$$

Write the decimal point in the product.

$$\begin{array}{r} 0.02 \\ \times\ 0.3 \\ \hline 0.006 \end{array}$$

0.02 ← 2 decimal places
× 0.3 ← 1 decimal place
0.006 ← 3 decimal places

Write 2 zeros to the left of 6.

Write the decimal point in the product. Write in zeros where necessary.

1. $\begin{array}{r} 0.3 \\ \times\ 0.7 \\ \hline 21 \end{array}$ **2.** $\begin{array}{r} 0.2 \\ \times\ 0.4 \\ \hline 8 \end{array}$ **3.** $\begin{array}{r} 0.43 \\ \times\ 0.5 \\ \hline 215 \end{array}$ **4.** $\begin{array}{r} 0.41 \\ \times\ 0.3 \\ \hline 123 \end{array}$ **5.** $\begin{array}{r} 0.05 \\ \times\ 0.1 \\ \hline 5 \end{array}$

6. $\begin{array}{r} 0.53 \\ \times\ 0.4 \\ \hline 212 \end{array}$ **7.** $\begin{array}{r} 0.23 \\ \times\ 0.3 \\ \hline 69 \end{array}$ **8.** $\begin{array}{r} 0.05 \\ \times\ 0.8 \\ \hline 40 \end{array}$ **9.** $\begin{array}{r} 0.02 \\ \times\ 0.6 \\ \hline 12 \end{array}$ **10.** $\begin{array}{r} 0.04 \\ \times\ 0.4 \\ \hline 16 \end{array}$

Find the product.

11. $8 \times 0.002 =$ ______ **12.** $0.1 \times 0.51 =$ ______ **13.** $0.2 \times 0.32 =$ ______

14. $0.3 \times 0.05 =$ ______ **15.** $0.01 \times 0.5 =$ ______ **16.** $0.02 \times 0.4 =$ ______

17. $1.2 \times 0.02 =$ ______ **18.** $6.1 \times 0.01 =$ ______ **19.** $0.02 \times 0.3 =$ ______

20. $4 \times 0.002 =$ ______ **21.** $4.1 \times 0.02 =$ ______ **22.** $0.5 \times 0.07 =$ ______

Circle the letter of the correct answer.

23. 6×0.005 **a.** 0.3 **b.** 0.03 **c.** 0.003

24. 7×0.004 **a.** 0.028 **b.** 0.28 **c.** 2.8

25. 0.08×0.5 **a.** 0.4 **b.** 0.04 **c.** 0.004

PROBLEM SOLVING

26. Don has 6 test tubes each containing 0.01 L of a glucose solution. How much glucose solution does he have? ______________

27. A piece of wire measured 0.03 m. Jane cut off 0.3 of the wire. How long was the piece she cut off? ______________

Dividing by 10, 100, and 1,000

Name ______________________

Date ______________________

Divide:

4.2 ÷ 10 = ?	4.2 ÷ **10** = 0.42	← 1 zero. Move 1 places to the left.
12.5 ÷ 100 = ?	12.5 ÷ **100** = 0.125	← 2 zeros. Move 2 places to the left.
174 ÷ 1000 = ?	174 ÷ **1000** = 0.174	← 3 zeros. Move 3 places to the left.
7.1 ÷ 100 = ?	07.1 ÷ **100** = 0.071	← 2 zeros. Move 2 places to the left. Write 1 zero as a place holder.

Write the quotient.

1. 521.6 ÷ 10 = ______
521.6 ÷ 100 = ______

2. 647 ÷ 10 = ______
647 ÷ 100 = ______

3. 73.9 ÷ 10 = ______
73.9 ÷ 100 = ______

4. 1825 ÷ 10 = ______
1825 ÷ 100 = ______
1825 ÷ 1000 = ______

5. 88 ÷ 10 = ______
88 ÷ 100 = ______
88 ÷ 1000 = ______

6. 4 ÷ 10 = ______
4 ÷ 100 = ______
4 ÷ 1000 = ______

Divide.

7. 0.6 ÷ 10 = ______
8. 0.03 ÷ 10 = ______
9. 7 ÷ 10 = ______
10. 5.32 ÷ 10 = ______
11. 0.9 ÷ 100 = ______
12. 4 ÷ 100 = ______
13. 16.7 ÷ 100 = ______
14. 0.36 ÷ 10 = ______
15. 4781.9 ÷ 100 = ______
16. 2 ÷ 1000 = ______
17. 648 ÷ 1000 = ______
18. 45,612 ÷ 1000 = ______

Find the missing numbers.

19. 5.06 ÷ ______ = 0.506
20. 7.3 ÷ ______ = 0.073
21. 78 ÷ ______ = 0.078
22. 841.2 ÷ ______ = 8.412
23. ______ ÷ 10 = 0.085
24. ______ ÷ 100 = 0.093
25. ______ ÷ 1000 = 4.125
26. ______ ÷ 100 = 0.68
27. ______ ÷ 1000 = 0.012

PROBLEM SOLVING

28. Howard cut 658.4 cm of rope into 100 equal pieces. How long is each piece? ______

29. A sporting goods store paid $19,950 for 1000 sweatshirts. How much did each sweatshirt cost? ______

Dividing Decimals by Whole Numbers

Name ________________________

Date ________________________

Divide: 3.76 ÷ 4 = ?

Write the decimal point of the quotient above the decimal point of the dividend.

4)3.76

Divide as you would with whole numbers.

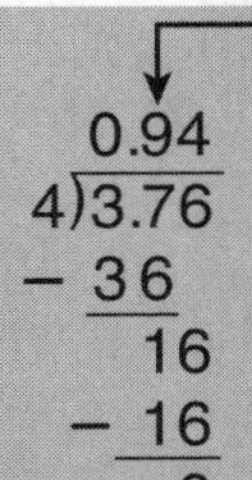

4 > 3 **Not enough** ones
4 < 37 **Enough** tenths
The quotient begins in the tenths place.

Check.

0.94
× 4
3.76

Circle the letter of the correct answer.

1. 3)3.6 **a.** 1.2 **b.** 12 **c.** 0.12 **d.** 0.012
2. 5)0.55 **a.** 1.1 **b.** 11 **c.** 0.11 **d.** 0.011
3. 6)0.84 **a.** 0.14 **b.** 1.4 **c.** 0.014 **d.** 14
4. 3)0.753 **a.** 25.1 **b.** 2.51 **c.** 0.251 **d.** 0.0251

Divide and check.

5. 4)1.6
6. 7)3.5
7. 9)8.1
8. 7)6.3
9. 5)0.70
10. 2)0.38
11. 3)0.54
12. 6)0.96
13. 6)$.78
14. 5)$9.45
15. 6)$21.72
16. 8)$77.36
17. 4.2 ÷ 7 = ______
18. 11.8 ÷ 2 = ______
19. $.87 ÷ 3 = ______
20. 2.59 ÷ 7 = ______
21. 49.38 ÷ 6 = ______
22. $25.35 ÷ 3 = ______
23. $58.48 ÷ 8 = ______
24. $18.99 ÷ 9 = ______
25. $8.12 ÷ 7 = ______

PROBLEM SOLVING

26. If Carmen paid $32.70 for 5 rolls of film, how much did she pay for each roll? ______________

Zeros in Division

Name ______________________________

Date ______________________________

Divide: 0.215 ÷ 5 = ?

Divide. **Check.**

Write the decimal point in the quotient.

$5\overline{)0.215}$

$5\overline{)0.21^{1}5}$ with quotient 0.043

5 > 2 **Not enough** tenths
Write 0 in the tenths place.
5 < 21 **Enough** hundredths
The quotient begins in the hundredths place.

$$\begin{array}{r} 0.043 \\ \times \quad 5 \\ \hline 0.215 \end{array}$$

Divide and check.

1. $6\overline{)0.024}$
2. $3\overline{)0.006}$
3. $7\overline{)7.21}$
4. $9\overline{)36.27}$
5. $4\overline{)8.12}$
6. $3\overline{)2.118}$
7. $4\overline{)4.12}$
8. $6\overline{)0.018}$
9. $5\overline{)0.8}$
10. $8\overline{)0.60}$
11. $4\overline{)4.2}$
12. $6\overline{)9.018}$
13. $2\overline{)3.016}$
14. $7\overline{)7.035}$
15. $9\overline{)18.027}$
16. $5\overline{)5.01}$

Divide. Round the quotient to the nearest thousandth.

17. $3\overline{)8.5}$
18. $7\overline{)3.2}$
19. $6\overline{)5.3}$
20. $3\overline{)9.44}$
21. $9\overline{)0.53}$
22. $7\overline{)0.81}$
23. $9\overline{)0.93}$
24. $6\overline{)0.82}$

 Use with Lesson 11-8, text pages 366–367.

Estimating Decimal Quotients

Name ______________________

Date ______________________

Estimate: 251.72 ÷ 8

$$8\overline{)251.72}\quad 30.00$$

8 > 2 **Not enough** hundreds
8 < 25 **Enough** tens
The quotient begins in the tens place.
About how many 8s in 25? **3**

The quotient is greater than 30.

Estimate: 4.53 ÷ 6

Use compatible numbers.

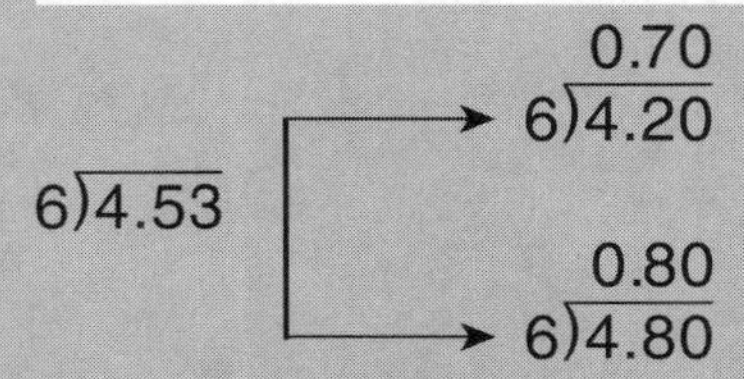

The quotient is between 0.70 and 0.80.

Estimate the quotient.

1. $5\overline{)0.156}$
2. $8\overline{)0.651}$
3. $4\overline{)2.542}$
4. $7\overline{)5.125}$
5. $6\overline{)13.025}$
6. $5\overline{)45.834}$
7. $8\overline{)10.728}$
8. $9\overline{)9.381}$
9. $3\overline{)0.71}$
10. $2\overline{)7.306}$
11. $6\overline{)26.535}$
12. $9\overline{)67.733}$
13. $8\overline{)34.655}$
14. $9\overline{)594.312}$
15. $7\overline{)623.118}$
16. $3\overline{)117.459}$

Estimate the quotient. Use compatible numbers.

17. $6\overline{)4.765}$
18. $8\overline{)7.196}$
19. $7\overline{)5.528}$
20. $4\overline{)29.001}$
21. $52\overline{)3.162}$
22. $68\overline{)4.305}$
23. $4\overline{)13.25}$
24. $6\overline{)3.55}$
25. $7\overline{)1.52}$
26. $91\overline{)28.13}$
27. $78\overline{)63.90}$
28. $32\overline{)152.68}$

PROBLEM SOLVING

29. Ms. Yakimoto drove 354.046 km in 4 hours. About how many kilometers did she drive in one hour? ______________

30. There are 589.7 mL of water evenly divided into 222 test tubes. About how much water is in each test tube? ______________

Estimating Money

Name ______________________

Date ______________________

Estimate: \$25.56 ÷ 3

Use compatible numbers.

3)\$25.56 → 3)\$24.00 = \$ 8.00; 3)\$27.00 = \$ 9.00

The exact quotient is between \$8.00 and \$9.00.

Divide: \$4.34 ÷ 3 = ?

3)\$4.340 = \$1.446

6 > 5
Round **up** to \$1.45.

Add a zero in the dividend.

Write what compatible numbers you would use. Then estimate the quotient.

1. 4)\$2.17 ______ **2.** 8)\$7.25 ______ **3.** 2)\$1.49 ______ **4.** 5)\$3.35 ______

5. 52)\$98.35 ______ **6.** 16)\$74.39 ______ **7.** 33)\$68.98 ______ **8.** 25)\$78.16 ______

9. 14)\$165.27 ______ **10.** 28)\$346.64 ______ **11.** 65)\$296.50 ______ **12.** 58)\$835.12 ______

Divide. Round the quotient to the nearest cent.

13. 7)\$8.85 **14.** 9)\$8.70 **15.** 6)\$7.41 **16.** 5)\$4.92

17. 8)\$42.36 **18.** 4)\$27.41 **19.** 3)\$65.51 **20.** 7)\$22.09

PROBLEM SOLVING

21. Jamal bought 5 lb of onions for \$4.98. About how much did 1 lb of onions cost? ______

22. One dozen fine-point pens cost \$15.39 and 5 erasers cost \$2.39. To the nearest cent, what is the cost of 1 fine-point pen? what is the cost of 1 eraser? ______

Problem-Solving Strategy: Write a Number Sentence

Name ______________________

Date ______________________

Frank hiked 0.75 as far as his brother, Joe.
If Joe hiked 4.8 miles, how far did Frank hike?

Let d represent the unknown distance.

0.75×4.8 mi $= d$

0.75×4.8 mi $= 3.6$ mi; $d = 3.6$ mi

Frank hiked 3.6 miles.

Solve. Do your work on a separate sheet of paper.

1. Harry's cat weighs 8.75 lb, and Maria's cat weighs 3.65 lb more. How much does Maria's cat weigh?

   ____

2. Lisa can run 3 miles in 24.12 min. How long will it take her to run 1 mile if she runs each mile at the same speed?

   ____

3. Nina bought 4.3 lb of potato salad and 3.65 lb of macaroni salad for a picnic. How much more potato salad than macaroni salad did she buy?

   ____

4. The distance from Paul's house to the lake shore is 86.25 m. The distance from the lake shore to the raft is 3.8 m. What is the total distance from Paul's house to the raft?

   ____

5. At Oscar's Office Supply Center, one pen costs $1.19. How much do 15 of the same kind of pens cost?

   ____

6. Sam earns $5.88 per hour. Last week he worked 9.5 hours. How much did he earn?

   ____

7. Judy is half as old as Shawn. If Shawn is 23.5 years old, how old is Judy?

   ____

8. Mr. Watanabi drives 224.4 mi in 4 h. What is his rate of speed per hour?

   ____

9. Beth lifts weights at the gym. She lifts 5.5 lb on Monday and 7 lb on Tuesday. How much more does she lift on Tuesday than on Monday?

   ____

10. Marisol sister lives 144 miles away. If Marisol drives at a rate of 48 miles per hour, how long should it take her to drive to her sister's house?

    ____

Metric Measurement

Name ______________________

Date ______________________

Metric Units of Length	(10 × 100) m = 1 kilometer (km)	1 meter (m)	(1 ÷ 10) m = 1 decimeter (dm)	(1 ÷ 100) m = 1 centimeter (cm)	(1 ÷ 1000) m = 1 millimeter (mm)
Metric Units of Capacity	(10 × 100) L = 1 kiloliter (kL)	1 liter (L)	(1 ÷ 10) L = 1 deciliter (dL)	(1 ÷ 100) L = 1 centiliter (cL)	(1 ÷ 1000) L = 1 milliliter (mL)
Metric Units of Mass	(10 × 100) g = 1 kilogram (kg)	1 gram (g)	(1 ÷ 10) g = 1 decigram (dg)	(1 ÷ 100) g = 1 centigram (cg)	(1 ÷ 1000) g = 1 milligram (mg)

Which metric unit is used to measure each? Write *m, L,* or *g*.

1. distance of a race ________
2. mass of a banana ________
3. capacity of a fish tank ________
4. height of a fence ________
5. length of a car ________
6. mass of a barrel ________
7. water in a pool ________
8. perimeter of a house ________
9. mass of a grape ________
10. milk in a bottle ________

Write the letter of the larger unit of measure.

11. _____ **a.** kiloliter **b.** deciliter
12. _____ **a.** milligram **b.** gram
13. _____ **a.** decimeter **b.** centimeter
14. _____ **a.** liter **b.** milliliter
15. _____ **a.** meter **b.** kilometer
16. _____ **a.** decigram **b.** kilogram

Write the letter of the smaller unit of measure.

17. _____ **a.** centigram **b.** gram
18. _____ **a.** centimeter **b.** millimeter
19. _____ **a.** centiliter **b.** liter
20. _____ **a.** kilometer **b.** decimeter
21. _____ **a.** kilogram **b.** gram
22. _____ **a.** liter **b.** kiloliter

Write the letter of the best metric unit to be used for each.

23. mass of a baseball **a.** kilogram **b.** gram **c.** milligram
24. distance to the moon **a.** kilometer **b.** meter **c.** decimeter
25. water in a spoon **a.** kiloliter **b.** liter **c.** milliliter
26. height of a building **a.** meter **b.** decimeter **c.** centimeter
27. mass of a truck **a.** kilogram **b.** decigram **c.** milligram
28. juice in a glass **a.** kiloliter **b.** liter **c.** centiliter

 Use with Lesson 12-1, text pages 382–383.

Renaming Metric Units

Name ______________________

Date ______________________

Multiply to rename larger units as smaller units.	Divide to rename smaller units as larger units.
3.7 m = ? cm	800 g = ? kg
3.7 m = (3.7 × 100) cm	800 g = (800 ÷ 1000) kg
3.7 m = 370 cm	800 g = 0.8 kg

Write the letter of the correct answer.

		a.	b.	c.	d.
1.	18 cm = ____ mm	1.8	18	180	1800
2.	10 kg = ____ g	1.0	100	1000	10 000
3.	6.7 g = ____ mg	6700	0.067	0.67	0.0067
4.	72.5 dm = ____ m	0.0725	7.25	0.725	7250

Complete.

5. 372 cg = ______ g
6. 348 g = ______ kg
7. 80 dL = ______ L
8. 18 kg = ______ g
9. 120 dL = ______ L
10. 60 cm = ______ dm
11. 642 mm = ______ m
12. 60 600 L = ______ kL
13. 2500 m = ______ km
14. 60 L = ______ cL
15. 900 mm = ______ cm
16. 2 m = ______ cm
17. 0.834 km = ______ m
18. 7.29 cm = ______ mm
19. 13.3 dm = ______ m
20. 2900 cg = ______ g
21. 11 m = ______ mm
22. 37 200 m = ______ km

PROBLEM SOLVING

23. Blueberry rakers filled 50 boxes that each had a mass of 2000 g. How many kilograms of berries were gathered?

____

24. Katrina needed to dilute 0.07 L of sugar solution with an equal amount of water for a science experiment. How many milliliters of water did she add?

____

25. Marty placed 10 dimes next to each other in a row. Each dime was 18 mm wide. How long was the row of coins in millimeters? in centimeters?

____

26. The distance between 2 cities is 630 km. A plane left one city and flew toward the other city. If it has flown 60 km 3000 m, has it reached its destination? If not, how much farther must the plane travel?

____

Relating Metric Units of Length

Name ______________________

Date ______________________

The length of the piece of yarn is
- 1 dm to the nearest decimeter.
- 8 cm to the nearest centimeter.
- 79 mm to the nearest millimeter.

Circle the letter of the unit you would use to measure each.

1. width of a calculator	**a.**	kilometer	**b.**	decimeter	**c.**	centimeter
2. height of a wall	**a.**	kilometer	**b.**	meter	**c.**	decimeter
3. thickness of a quarter	**a.**	decimeter	**b.**	centimeter	**c.**	millimeter
4. height of a person	**a.**	meter	**b.**	decimeter	**c.**	millimeter
5. distance to Mars	**a.**	kilometer	**b.**	meter	**c.**	decimeter

Write the name of an item you would measure in:

6. millimeters ______________ **7.** centimeters ______________

8. decimeters ______________ **9.** meters ______________

10. kilometers ______________________

Measure each to the nearest millimeter, nearest centimeter, and nearest decimeter.

11. ______ mm ______ cm ______ dm

12. ______ mm ______ cm ______ dm

13.

______ mm ______ cm ______ dm

14.

______ mm ______ cm ______ dm

PROBLEM SOLVING

15. Suppose you are describing the size of your math workbook. What unit would you use? Why?

____

Relating Metric Units of Capacity and Mass

Name ______________________________

Date ______________________________

1 L = 1000 mL	1 g = 1000 mg
1 L = 100 cL	1 g = 100 cg
1 L = 10 dL	1 g = 10 dg
1 kL = 1000 L	1 kg = 1000 g
	1 t = 1000 kg

Circle the letter of the unit you would use to measure the capacity of each.

1. swimming pool	**a.** kiloliter	**b.** liter	**c.** deciliter
2. aquarium	**a.** liter	**b.** centiliter	**c.** milliliter
3. soup bowl	**a.** kiloliter	**b.** liter	**c.** milliliter
4. bucket	**a.** liter	**b.** deciliter	**c.** milliliter

Circle the letter of the unit you would use to measure the mass of each.

5. baseball card	**a.** kilogram	**b.** gram	**c.** milligram
6. human being	**a.** metric ton	**b.** kilogram	**c.** milligram
7. computer	**a.** kilogram	**b.** gram	**c.** decigram
8. Earth	**a.** metric ton	**b.** kilogram	**c.** gram
9. crayon	**a.** kilogram	**b.** gram	**c.** milligram

Compare. Write <, =, or >.

10. 3 L ____ 300 mL	**11.** 16 L ____ 15 000 mL	**12.** 500 cL ____ 6L
13. 4 kL ____ 4000 L	**14.** 670 dL ____ 68 L	**15.** 4000 L ____ 40 000 kL
16. 460 mL ____ 4.6 L	**17.** 5000 L ____ 5 kL	**18.** 2 L ____ 20 dL
19. 182 L ____ 18 200 cL	**20.** 0.8 L ____ 9 dL	**21.** 0.7 L ____ 650 mL
22. 42 cL ____ 4 L	**23.** 18 000 mL ____ 18 L	**24.** 8.9 L ____ 88 dL
25. 2 kg ____ 1500 g	**26.** 3 kg ____ 30 000 g	**27.** 2300 kg ____ 2.5 g
28. 7000 g ____ 70 kg	**29.** 1000 cg ____ 11 g	**30.** 6000 dg ____ 60 g
31. 8 g ____ 80 kg	**32.** 3000 kg ____ 3 t	**33.** 16 000 g ____ 15 kg
34. 300 mg ____ 3 dg	**35.** 5 g ____ 500 mg	**36.** 1 t ____ 1010 kg

PROBLEM SOLVING

37. Rose has 3 kg of turkey slices from which to make 20 sandwiches. If she divides the slices equally, how many grams of turkey will be in each sandwich? ______________________

 Use with Lessons 12-4 and 12-5, text pages 388–391.

Square Measure

Name ______________________

Date ______________________

Count the squares to find the area.

1 m² ➤

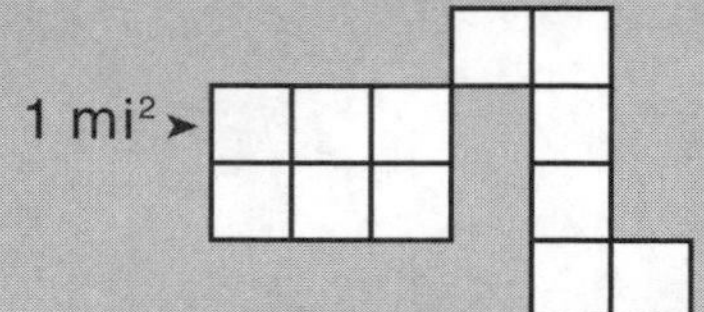

Each square = 1 m^2
Area = 12 m^2

1 mi² ➤

Each square = 1 mi^2
Area = 12 mi^2

Find the area of each figure.

1. 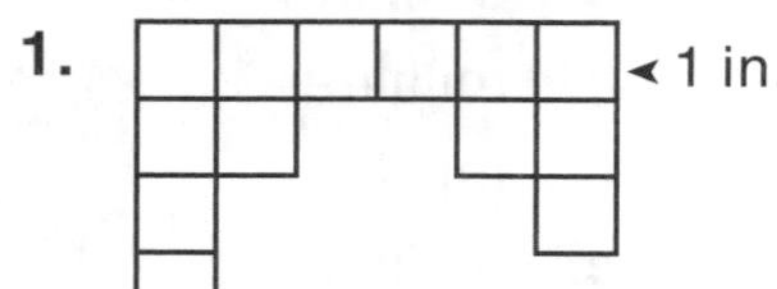  ◂ 1 $in.^2$

_______ $in.^2$

2. 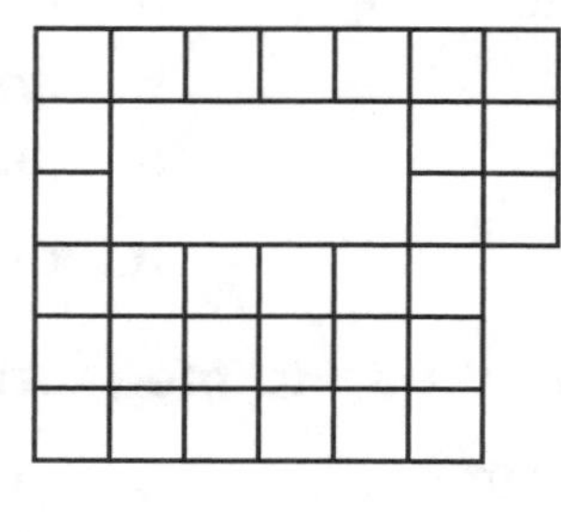 ◂ 1 yd^2

_______ yd^2

3. 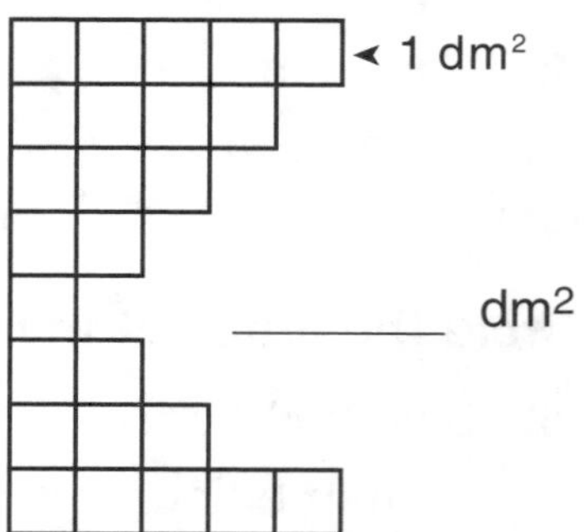 ◂ 1 dm^2

_______ dm^2

4. 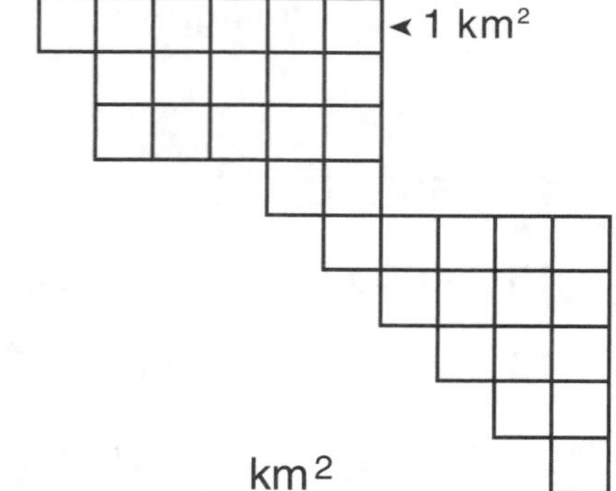 ◂ 1 km^2

_______ km^2

5. 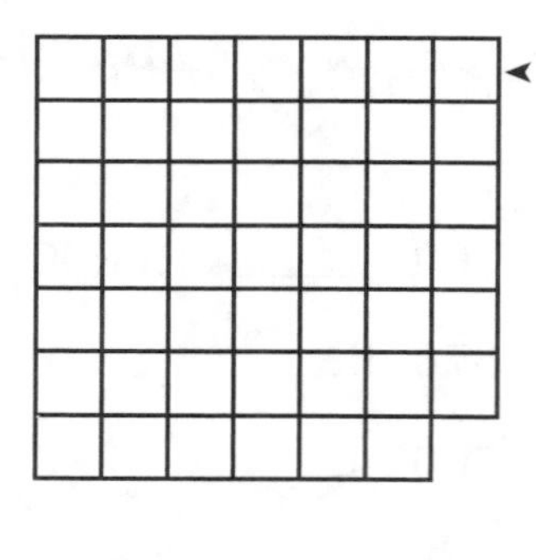 ◂ 1 ft^2

_______ ft^2

6.  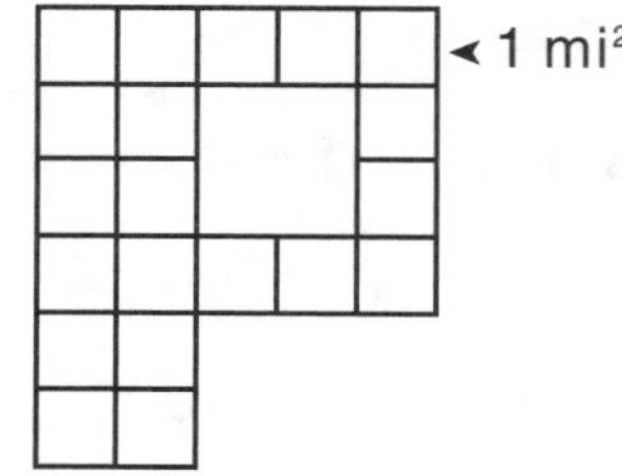 ◂ 1 mi^2

_______ mi^2

Estimate the area of each shaded figure.

7. km^2

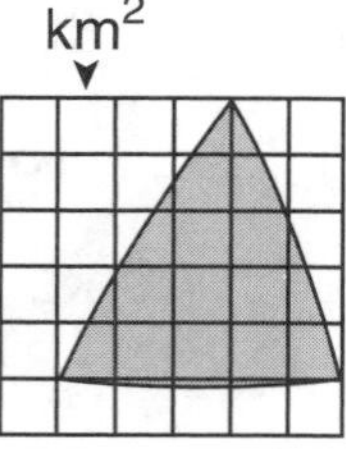

_______ km^2

8. yd^2

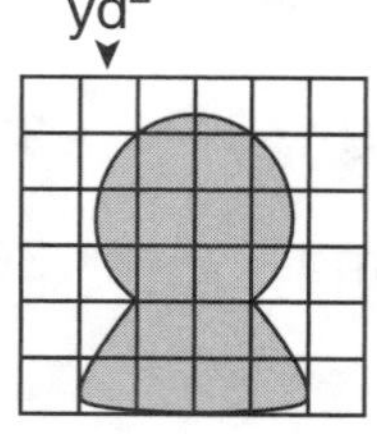

_______ yd^2

9. dm^2

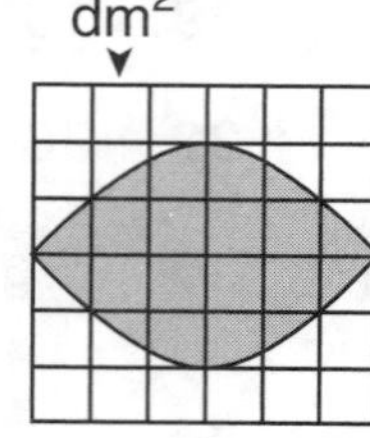

_______ dm^2

10. $in.^2$

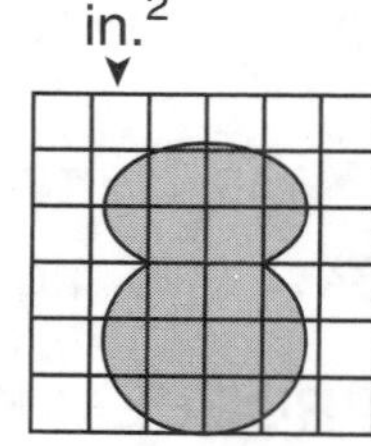

_______ $in.^2$

 Use with Lesson 12-6, text pages 392–393.

Areas of Rectangles and Squares

Name ______________________

Date ______________________

Rectangles

$A = \ell \times w$

$A = 17 \text{ ft} \times 10 \text{ ft}$

$A = 170 \text{ ft}^2$

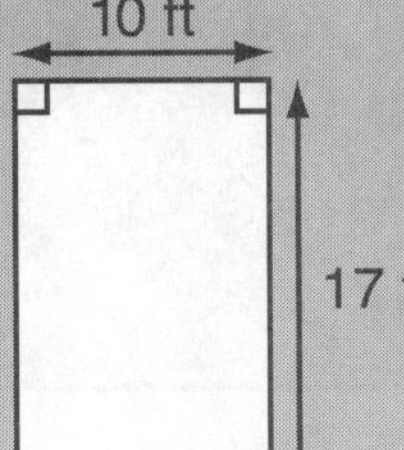

Squares

$A = s \times s$

$A = 12.3 \text{ m} \times 12.3 \text{ m}$

$A = 151.29 \text{ m}^2$

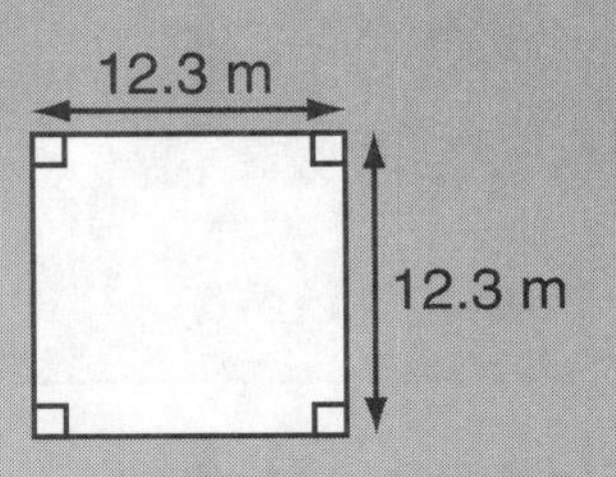

Find the area of each rectangle.

1.

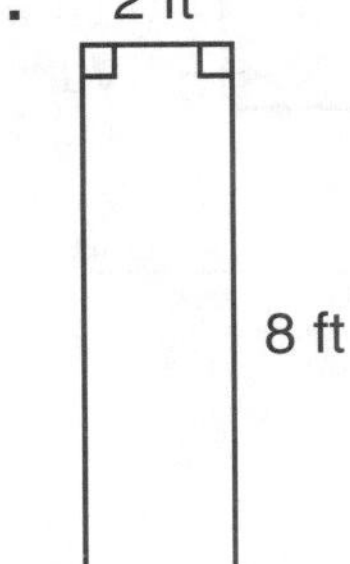

____

2.

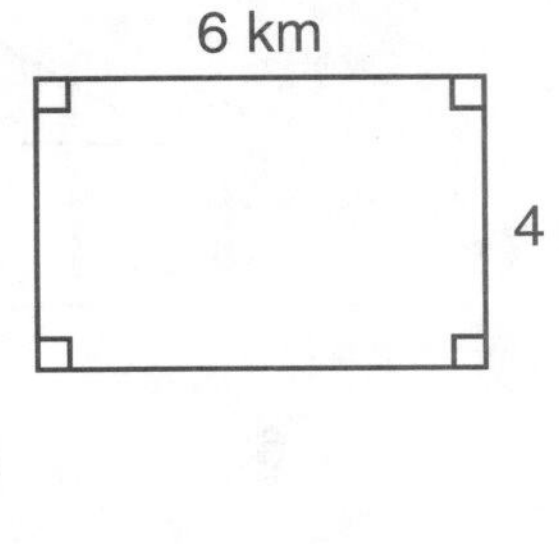

____

3.

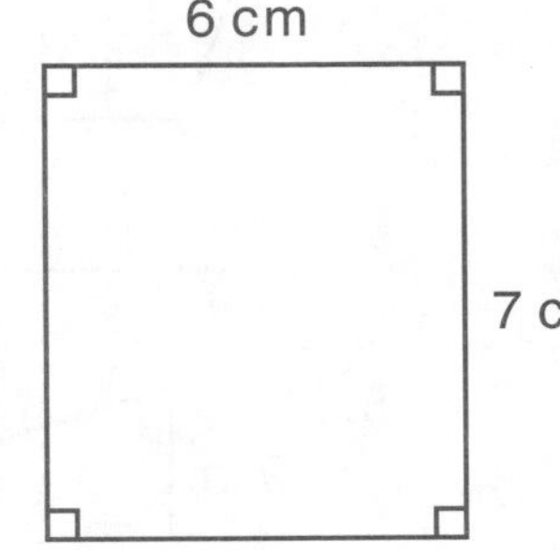

____

4.

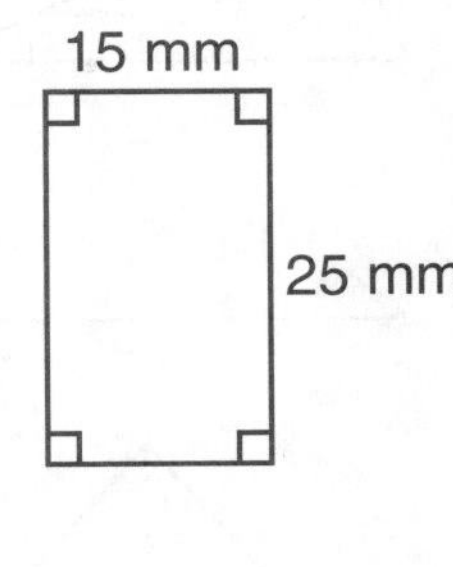

____

Find the area of each square.

5.

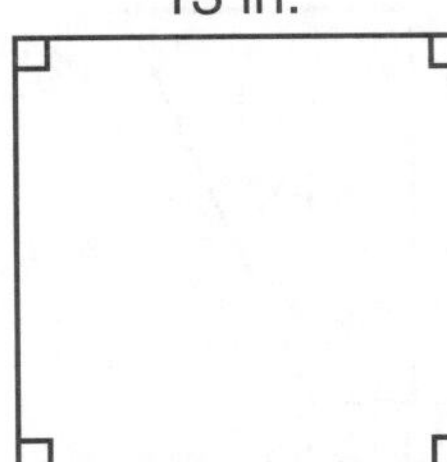

____

6.

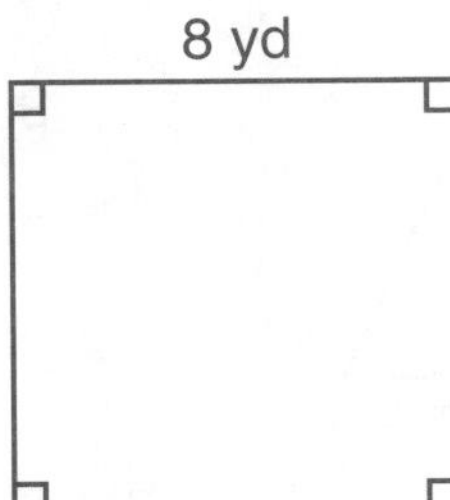

____

7.

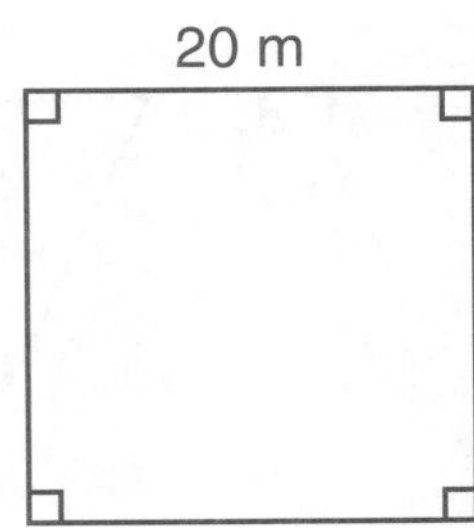

____

8.

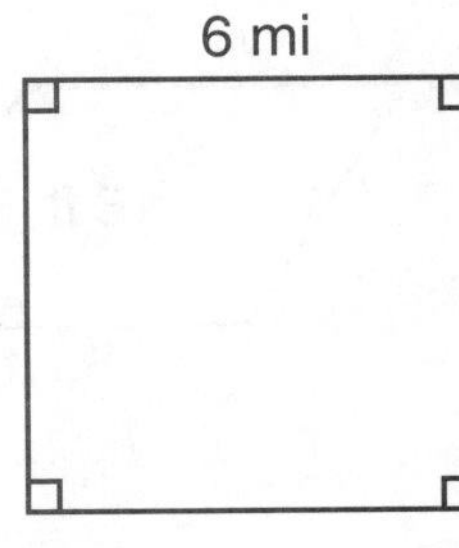

____

Find the area of each square or rectangle.

9. 7.3 dm long
3.9 dm wide

____

10. $s = 6\frac{1}{8}$ in.

____

11. 22 mm long
18 mm wide

____

12. $s = 4.9$ cm

____

13. 29 yd long
$9\frac{3}{4}$ yd wide

____

14. $s = 9.6$ m

____

15. 2 km long
0.5 km wide

____

16. $s = 8\frac{1}{2}$ ft

____

Areas of Parallelograms and Triangles

Name ______________________

Date ______________________

$A = \frac{1}{2} \times b \times h$

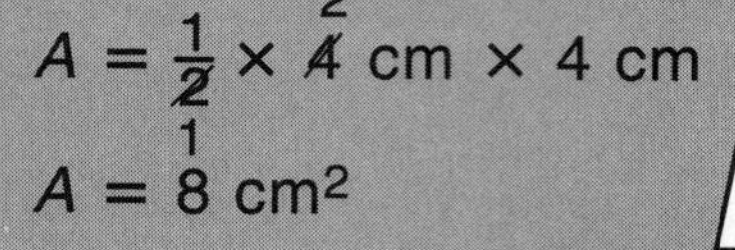

$A = \frac{1}{\not{2}_1} \times \not{4}^{2}\ \text{cm} \times 4\ \text{cm}$

$A = 8\ \text{cm}^2$

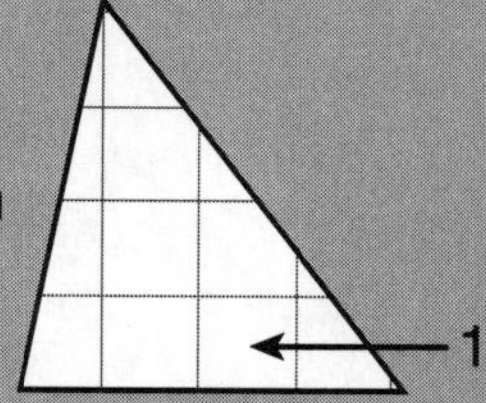

$A = b \times h$

$A = 5\ \text{in.} \times 3\ \text{in.}$

$A = 15\ \text{in.}^2$

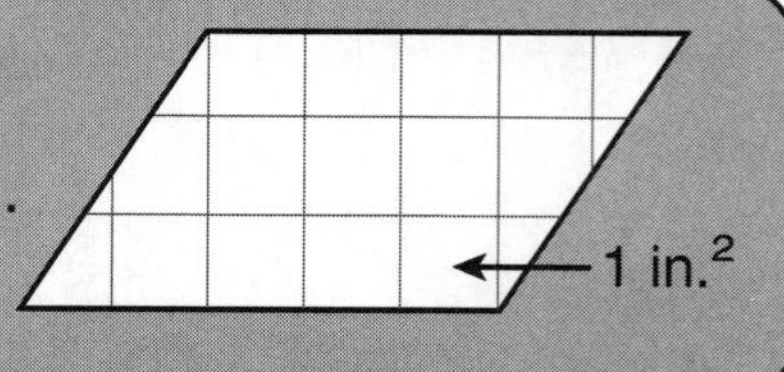

Find the area.

1.

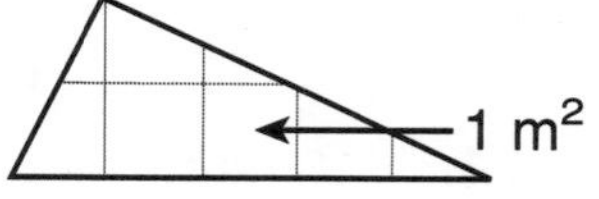

____

2.

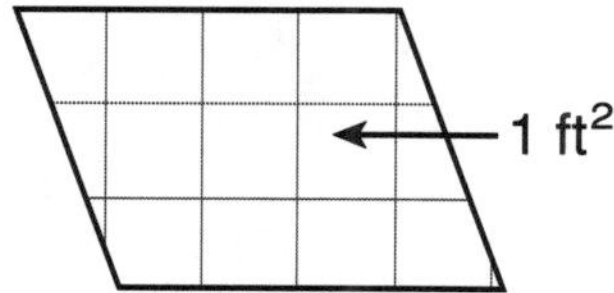

____

3.

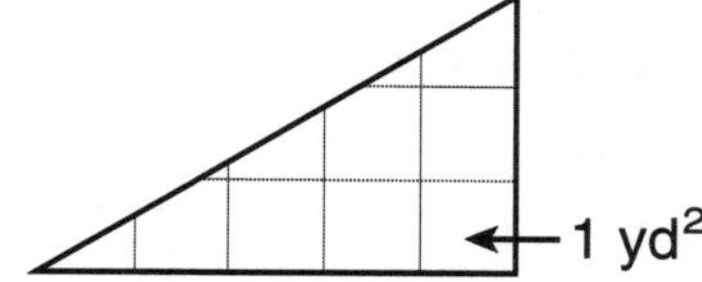

____

4.

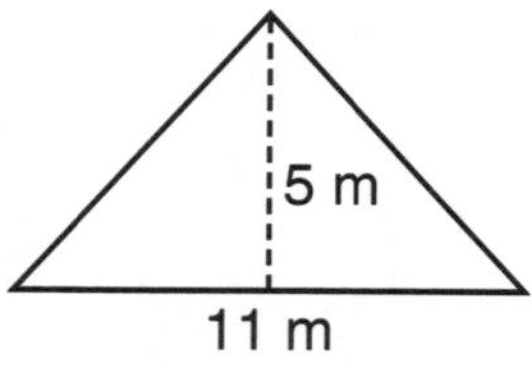

____

5.

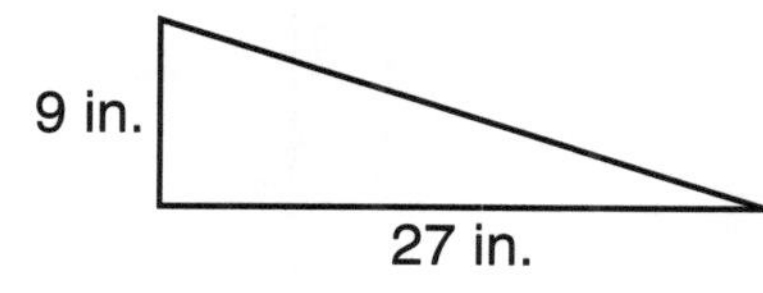

____

6.

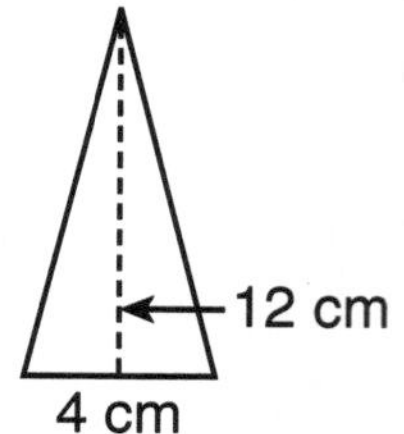

____

7.

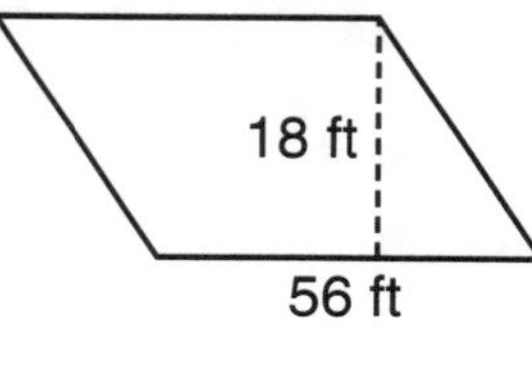

____

8.

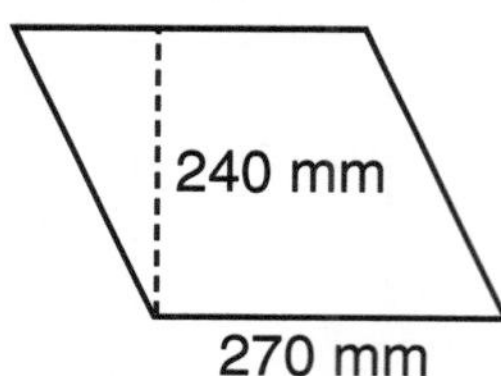

____

9.

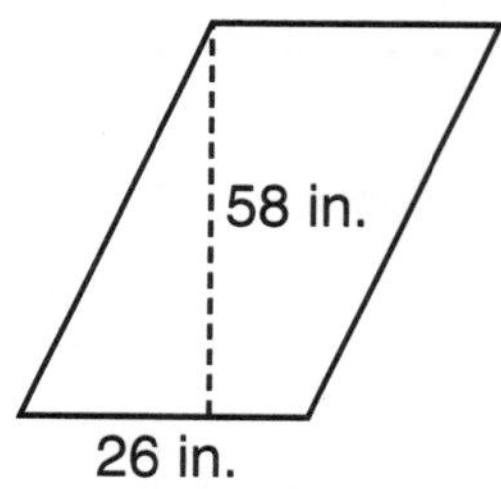

____

PROBLEM SOLVING

10. A square measures 14 in. along one side. A parallelogram has a base of 18 in. and a height of 10 in. Which figure has the greater area?

____

11. A parallelogram has a base of 32 cm and a height of 25 cm. A triangle has a base of 66 cm and a height of 25 cm. Which figure has the greater area? by how much?

____

Space Figures

Name ____________________

Date ____________________

base
base
vertex
face
edge

Pyramids and prisms other than cubes are named by the shape of their bases.

Cylinders, cones, and spheres have curved surfaces.

Complete the table.

	Space Figure	Number of Faces	Number of Vertices	Number of Edges
1.	square pyramid			
2.	rectangular pyramid			
3.	pentagonal prism			
4.	triangular pyramid			
5.	rectangular prism			
6.	pentagonal pyramid			
7.	hexagonal prism			
8.	triangular prism			
9.	hexagonal pyramid			

Write *True* or *False* for each statement.

10. A cone has no flat surfaces. ________

11. A sphere has one vertex. ________

12. A triangular prism has 9 edges. ________

13. A pentagonal pyramid has five edges. ________

14. A cylinder has a curved surface and 2 flat faces. ________

15. A cube is a special type of prism. ________

16. A square pyramid and a rectangular pyramid have the same number of faces and edges. ________

17. The base of a cube is a square. ________

18. A hexagonal prism has 8 hexagonal faces. ________

19. A triangular pyramid has 4 triangular faces. ________

20. Three of the faces of a triangular prism are rectangles. ________

21. The number of edges on a cube is triple the number of its faces. ________

 Use with Lesson 12-9, text pages 398–399.

Cubic Measure

Name ______________________

Date ______________________

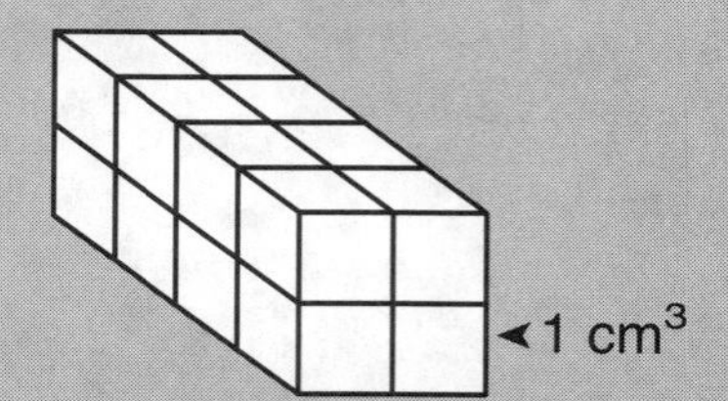

Count cubes to find the cubic measure of a space figure.

The cubic measure of the figure at the left is 16 cm^3.

Find the cubic measure in cubic units.

1. 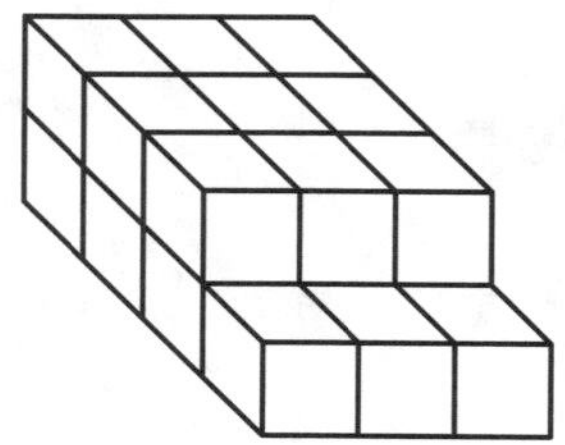

____

2. 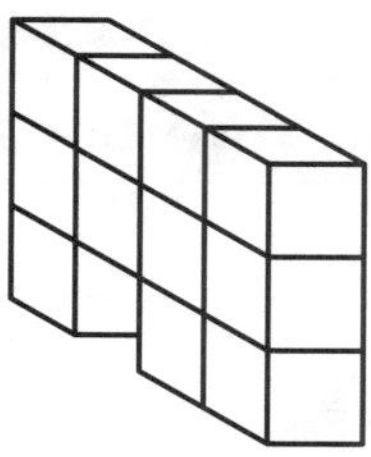

____

3. 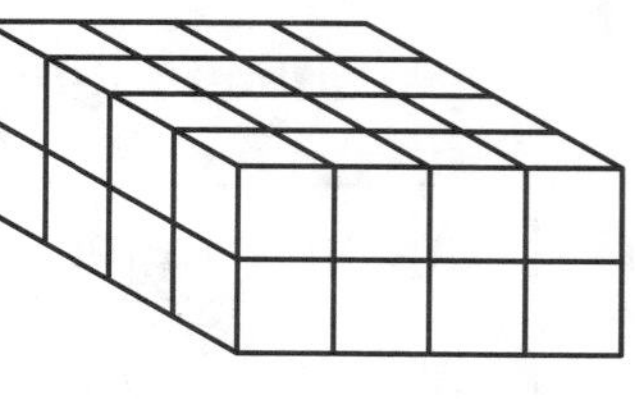

____

Find the cubic measure of each.

4. 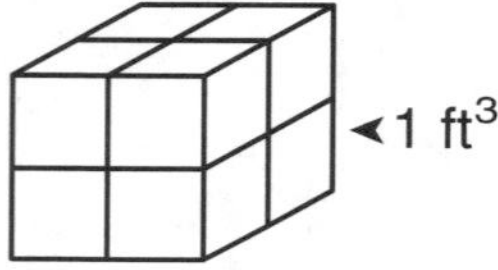

______ ft^3

5. 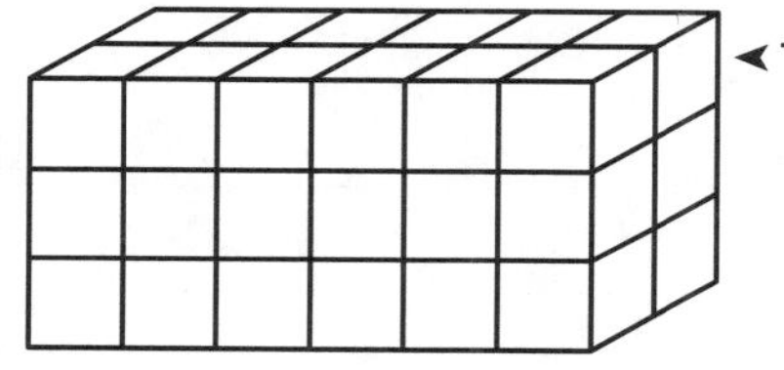

______ $in.^3$

6. 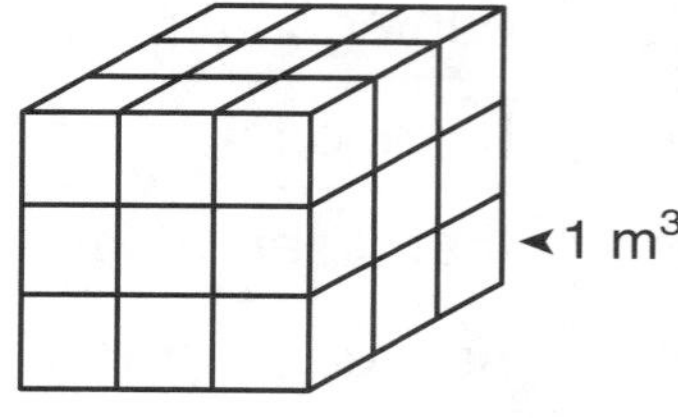

______ $in.^3$

7. 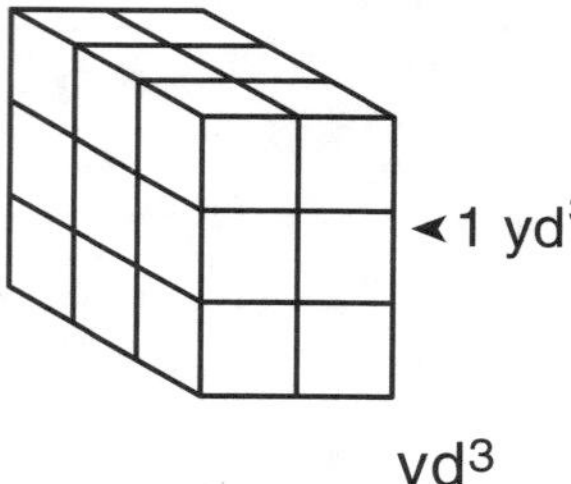

______ yd^3

8. 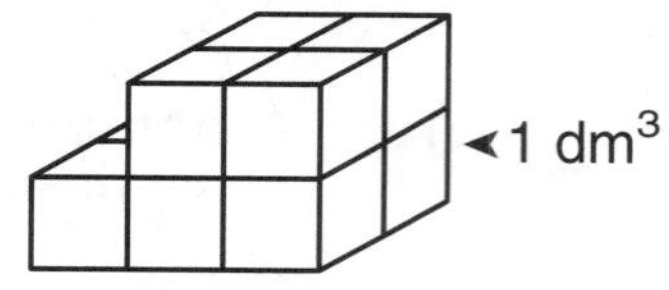

______ dm^3

9. 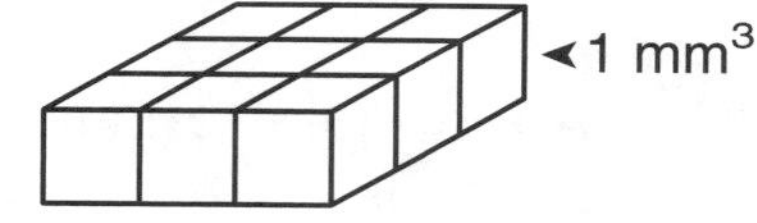

______ mm^2

Complete the table.

	Cubic Measure	Capacity of Water	Mass of Water
10.	2 cm^3	2 mL	
11.	4 dm^3		4 kg
12.		6 mL	6 g
13.	8 dm^3		
14.	10 cm^3		

 Use with Lesson 12-10, text pages 400–401.

Volume

Name ______________________

Date ______________________

Find the volume of the space figure.

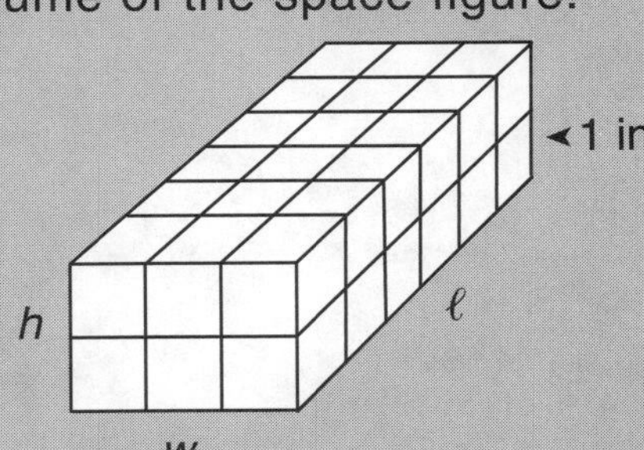

Use the formula.

$V = \ell \times w \times h$

$V = 6 \text{ in.} \times 3 \text{ in.} \times 2 \text{ in.}$

$V = 36 \text{ in.}^3$

Find the volume in cubic units.

1. 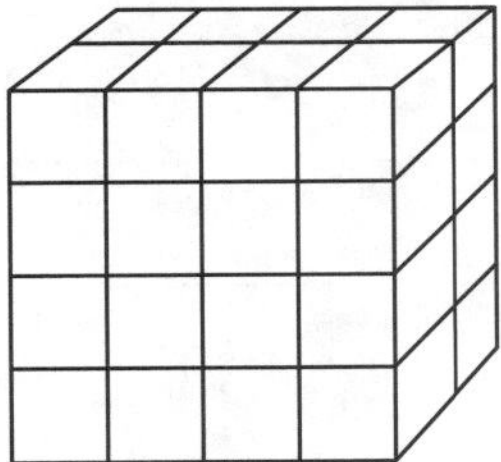 ______________

2. 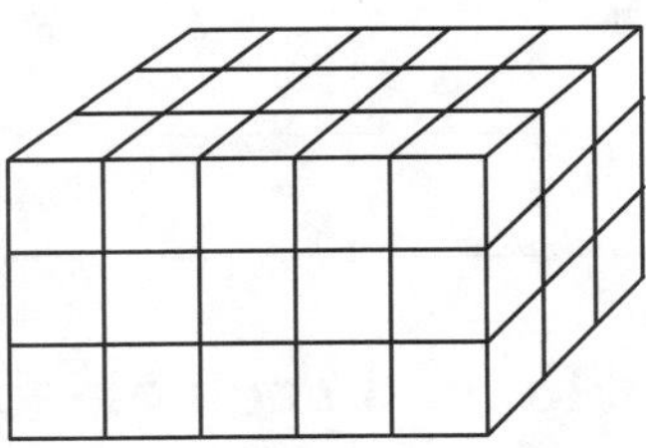 ______________

3. 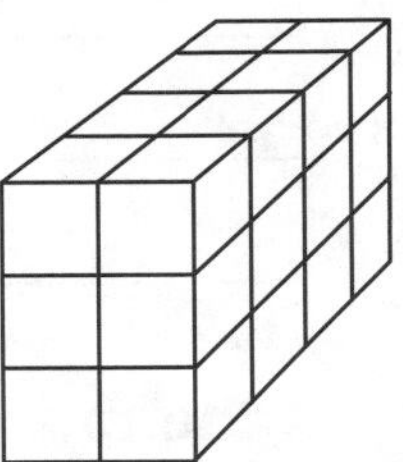 ______________

Find the volume of each rectangular prism.

4. 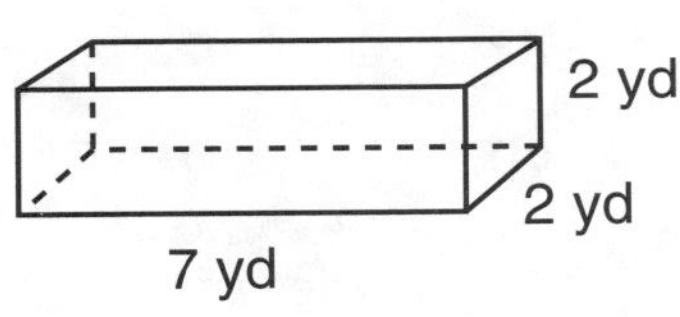

____

5. 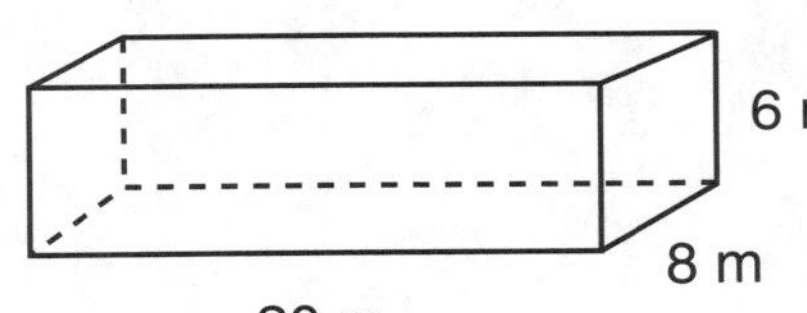

____

6. 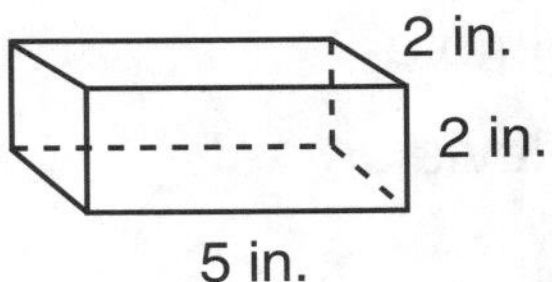

____

Complete the table.

	ℓ	w	h	$V = \ell \times w \times h$
7.	$21\frac{1}{2}$ ft	$3\frac{1}{3}$ ft	$2\frac{1}{2}$ ft	
8.	2.8 cm	2.4 cm	2.2 cm	
9.	8.3 dm	3.9 dm	4.8 dm	
10.	$29\frac{3}{4}$ in.	$10\frac{1}{2}$ in.	3 in.	
11.	6.4 m	5.2 m	12 m	

Estimating Volume

Name ______________________

Date ______________________

Net of a centimeter cube

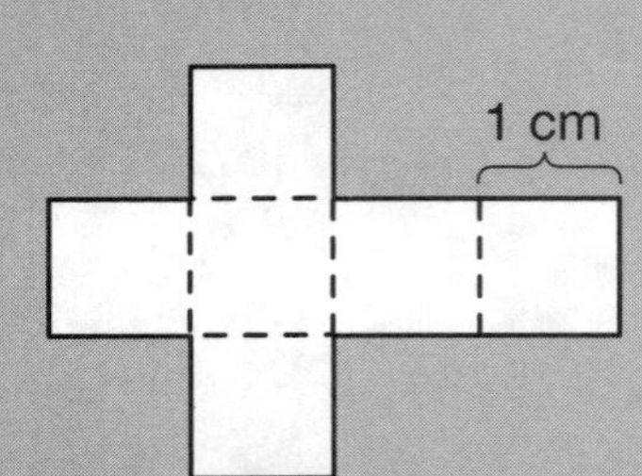

centimeter cube ($V = 1\ cm^3$)

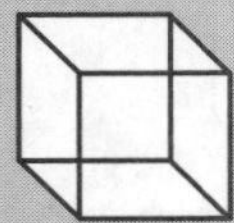

The shell would fit inside a centimeter cube.

Name 3 objects that would fit inside a cube with a volume of:

1. 1 cm^3 ______________________

2. 1 dm^3 ______________________

3. 1 $in.^3$ ______________________

Which size is the most reasonable to hold each object? Circle the letter of the best answer.

4.	mug	**a.** m^3	**b.** dm^3	**c.** cm^3	**d.** $in.^3$
5.	pearl	**a.** m^3	**b.** dm^3	**c.** cm^3	**d.** ft^3
6.	pitcher	**a.** $in.^3$	**b.** dm^3	**c.** cm^3	**d.** ft^3
7.	thimble	**a.** m^3	**b.** $in.^3$	**c.** ft^3	**d.** dm^3
8.	television set	**a.** m^3	**b.** dm^3	**c.** cm^3	**d.** ft^3

Circle the letter of the best estimate of volume for each.

9.	box for index cards	**a.** 60 yd^3	**b.** 60 ft^3	**c.** 60 $in.^3$
10.	cereal box	**a.** 3500 m^3	**b.** 3500 cm^3	**c.** 3500 mm^3
11.	board game box	**a.** 288 yd^3	**b.** 288 ft^3	**c.** 288 $in.^3$
12.	videocassette box	**a.** 660 dm^3	**b.** 660 cm^3	**c.** 660 mm^3
13.	refrigerator	**a.** 22 yd^3	**b.** 22 ft^3	**c.** 22 $in.^3$
14.	box for a computer monitor	**a.** 8505 dm^3	**b.** 8505 cm^3	**c.** 8505 mm^3

PROBLEM SOLVING

15. How many centimeter cubes fit into a decimeter cube? ______________________

16. How many inch cubes fit into a 1-foot cube? ______________________

Problem-Solving Strategy: Draw a Picture

Name ______________________

Date ______________________

Jane covered five faces of a box with colored paper, leaving the top open. The box is 4 cm high, 5.8 cm wide, and 6.5 cm long. How much paper did she use?

Draw a picture of the box.

top

4 cm

6.5 cm

5.8 cm

Find the area of each of the faces she covered.

Area = $\ell \times w$

Front: 4 cm × 5.8 cm = 23.2 cm^2

Back: 4 cm × 5.8 cm = 23.2 cm^2

Bottom: 5.8 cm × 6.5 cm = 37.7 cm^2

Right Side: 4 cm × 6.5 cm = 26 cm^2

Left Side: 4 cm × 6.5 cm = 26 cm^2

Add the areas: 23.2 cm^2 + 23.2 cm^2 + 26 cm^2 + 26 cm^2 + 37.7 cm^2 = 136.1 cm^2

Jane used 136.1 cm^2 of paper.

Solve. Do your work on a separate sheet of paper.

1. Laura pastes a blue right triangle in each corner of a red rectangle that is 8 cm wide and 12 cm long. The sides of the triangles are lined up with the sides of the rectangle, and none of the triangles overlap. Each triangle has a base of 4 cm and a height of 6 cm. How much red area is still showing?

   ____

2. Leona places a cube with a volume of 512 cm^3 into a rectangular fish tank that has two square faces, each with sides measuring 20 cm, and three rectangular faces each that are 40 cm long and 20 cm wide. How much of the total volume of the fish tank is left to be filled with water?

   ____

3. What is the least perimeter you can make by joining 4 rectangles side to side if each length is 6 cm and each width is 3.5 cm? What is the greatest perimeter?

   ____

4. A rectangle has a perimeter of 20 m and an area of 9 m^2. What are the length and width of the rectangle?

   ____

5. A rectangle has an area of 24 m^2. What are the possible pairs of lengths and widths? Which measurements give the greatest perimeter? Which measurements give the least perimeter?

   ____

6. Les made a rectangular path around his garden. The garden has a length of 18 ft and a width of 12 ft. If the outer edge of the path has a length of 22 ft and a width of 16 ft, how much area does the path cover?

   ____

Ratios as Fractions

Name ______________________

Date ______________________

The ratio of bees to butterflies is 2 to 3 or 2 : 3 or $\frac{2}{3}$.

The ratio of butterflies to bees is 3 to 2 or 3 : 2 or $\frac{3}{2}$.

$\frac{4}{6} = \underline{?}$

$\frac{4}{6} = \frac{4 \times 2}{6 \times 2} = \frac{8}{12}$

or

$\frac{4}{6} = \frac{4 \div 2}{6 \div 2} = \frac{2}{3}$

So $\frac{4}{6} = \frac{8}{12} = \frac{2}{3}$ are equal ratios.

There are 5 dogs, 6 cats, 7 birds, 11 mice, and 2 snakes in the pet store. Write each ratio in three ways.

1. dogs to cats ______
2. cats to birds ______
3. birds to mice ______
4. mice to snakes ______
5. snakes to dogs ______
6. birds to dogs ______
7. mice to cats ______
8. snakes to birds ______
9. dogs to mice ______
10. cats to mice ______

Write each ratio in simplest form.

11. 4 to 8 ______
12. 4 : 12 ______
13. $\frac{14}{7}$ ______
14. 16 to 32 ______
15. 24 to 6 ______
16. 7 : 18 ______
17. 16 : 24 ______
18. $\frac{15}{36}$ ______
19. $\frac{84}{16}$ ______
20. 6 to 36 ______
21. 13 : 100 ______
22. $\frac{16}{40}$ ______

Complete.

23. $\frac{1}{3} = \frac{\quad}{15}$
24. $\frac{2}{3} = \frac{\quad}{9}$
25. $\frac{4}{5} = \frac{\quad}{20}$
26. $\frac{3}{4} = \frac{\quad}{24}$
27. $\frac{12}{16} = \frac{\quad}{4}$
28. $\frac{4}{36} = \frac{\quad}{9}$
29. $\frac{8}{32} = \frac{\quad}{4}$
30. $\frac{9}{12} = \frac{\quad}{4}$
31. $\frac{3}{5} = \frac{\quad}{25}$
32. $\frac{21}{7} = \frac{\quad}{1}$
33. $\frac{110}{120} = \frac{\quad}{12}$
34. $\frac{3}{7} = \frac{\quad}{28}$

PROBLEM SOLVING

35. There are 18 girls and 12 boys in Ms. Lorenzo's class. What is the ratio in simplest form of boys to girls? ______

36. Tammy has 64 baseball cards and Ricardo has 100. What is the ratio in simplest form of the number of cards that Tammy has to the number of cards that Ricardo has? ______

Proportions

Name ____________________

Date ____________________

Do $\frac{1}{7}$ and $\frac{2}{14}$ form a proportion?	Find n: $\frac{1}{3} = \frac{n}{12}$
$\frac{1}{7} \overset{?}{=} \frac{2}{14}$ $\frac{1}{7} \times \frac{2}{14}$ $1 \times 14 = 7 \times 2$ $14 = 14$ So $\frac{1}{7} = \frac{2}{14}$ is a proportion.	$\frac{1}{3} = \frac{n}{12} \rightarrow \frac{1 \times 4}{3 \times 4} = \frac{4}{12}$ So $n = 4$.

Do the two given ratios form a proportion? Write *Yes* or *No*.

1. $\frac{5}{6}, \frac{15}{18}$ ______
2. $\frac{3}{4}, \frac{9}{16}$ ______
3. $\frac{2}{5}, \frac{5}{2}$ ______
4. $\frac{3}{7}, \frac{6}{14}$ ______
5. $\frac{1}{4}, \frac{3}{12}$ ______
6. $\frac{28}{21}, \frac{4}{3}$ ______
7. $\frac{18}{9}, \frac{9}{3}$ ______
8. $\frac{15}{20}, \frac{3}{4}$ ______

Use the cross-products rule to find out which of these are proportions. Write *Yes* or *No*.

9. $\frac{4}{10} = \frac{12}{30}$ ______
10. $\frac{1}{2} = \frac{8}{16}$ ______
11. $\frac{2}{3} = \frac{4}{9}$ ______
12. $\frac{3}{5} = \frac{9}{15}$ ______
13. $\frac{6}{12} = \frac{18}{36}$ ______
14. $\frac{18}{24} = \frac{6}{12}$ ______
15. $\frac{16}{20} = \frac{4}{5}$ ______
16. $\frac{6}{10} = \frac{18}{30}$ ______

Find the missing number in the proportion.

17. $\frac{2}{7} = \frac{4}{n}$ ______
18. $\frac{5}{6} = \frac{n}{18}$ ______
19. $\frac{3}{5} = \frac{n}{45}$ ______
20. $\frac{4}{n} = \frac{12}{60}$ ______
21. $\frac{1}{3} = \frac{1\frac{1}{3}}{n}$ ______
22. $\frac{1}{7} = \frac{1\frac{2}{7}}{n}$ ______
23. $\frac{n}{8} = \frac{1}{3}$ ______
24. $\frac{n}{7} = \frac{1}{5}$ ______
25. $\frac{\text{1 case}}{\text{4 cases}} = \frac{\text{12 bottles}}{n \text{ bottles}}$ ______
26. $\frac{\text{3 goldfish}}{\text{9 goldfish}} = \frac{\text{2 guppies}}{n \text{ guppies}}$ ______

PROBLEM SOLVING

27. Three goldfish cost \$2.00. How many goldfish will \$10.00 buy? ______
28. If 2 plums cost \$.28, how much do 12 plums cost? ______
29. If oranges cost \$2.40 a dozen, how much do 2 oranges cost? ______
30. If 2 cups of rice serve 5 people, how many cups of rice do you need to serve 60 people? ______

Scale and Maps

Name ______________________

Date ______________________

Complete the table. Measure the scale distance on the map to the nearest $\frac{1}{8}$ inch.

	Between Cities	Scale Distance (in.)	Actual Distance (mi)
1.	New Bedford-Rockville	$\frac{3}{4}$	
2.	New Bedford-Denton		
3.	Dover-Clinton		
4.	Dover-Weston		
5.	Rockville-Weston		

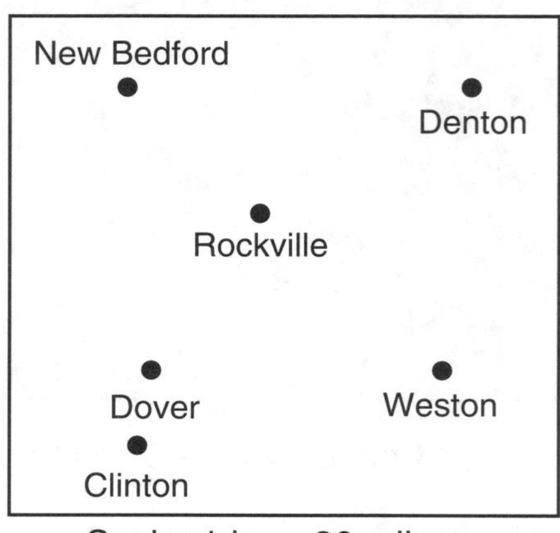

Use the scale 1 in. = 2 mi to complete each table.

	To go from:	Scale Distance	Actual Distance
6.	Troy to Dixmont	2 in.	
7.	Monroe to Swanville	$3\frac{1}{4}$ in.	
8.	Prospect Ferry to Sandy Point	$2\frac{1}{8}$ in.	

	To go from:	Scale Distance	Actual Distance
9.	Carmel to Etna	$1\frac{5}{8}$ in.	
10.	Carmel to Winterport	$7\frac{7}{8}$ in.	
11.	Dixmont to Brooks	$4\frac{1}{2}$ in.	

Measure the scale distance to the nearest centimeter. Then estimate each distance.

12. From Center City to Parville ____________
13. From Center City to Sikes Ave. ____________
14. The length of Long Ave. ____________
15. From road marker 12 to road marker 16 ____________
16. From road marker 9 to road marker 12 ____________
17. From Center City to road marker 16 ____________
18. From Center City to road marker 4 ____________
19. The shortest distance along a road from road marker 5 to road marker 4 ____________

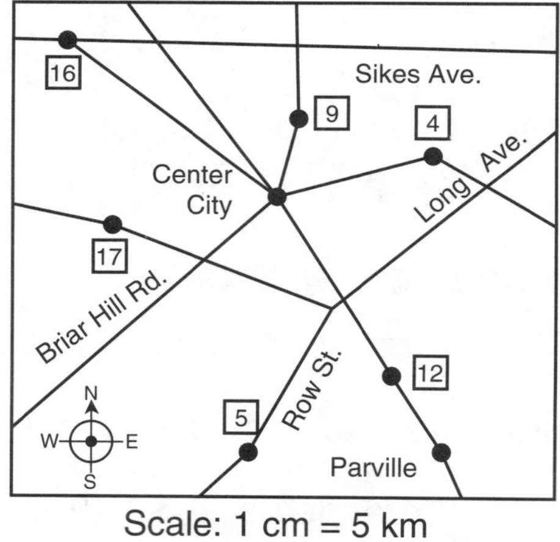

Scale: 1 cm = 5 km

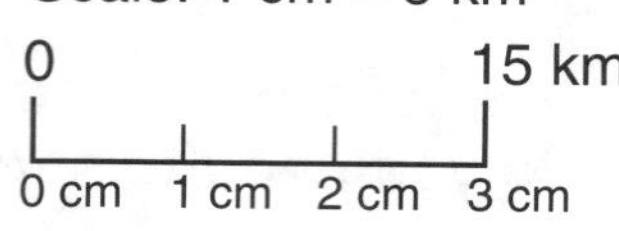

 Use with Lesson 13-3, text pages 420–421.

Relating Fractions to Percents

Name ____________________

Date ____________________

In the 100-square grid, 47 squares are shaded.

$\frac{47}{100}$ of the grid is shaded.

$\frac{47}{100} = 47\%$

So 47% of the grid is shaded.

Tell what fractional part of the grid is shaded. Then write the fraction as a percent.

1. 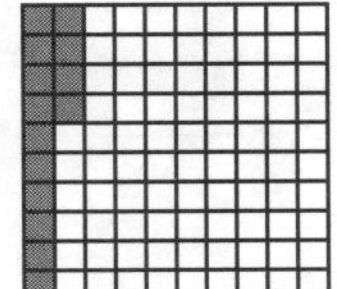 $\frac{\quad}{100}$ = ____%

2. 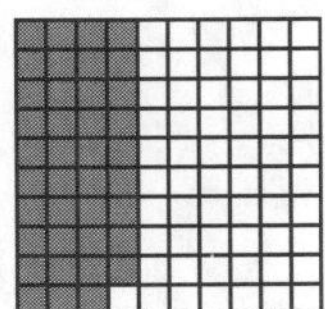 $\frac{\quad}{100}$ = ____%

3. 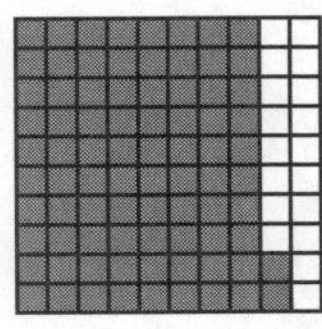 $\frac{\quad}{100}$ = ____%

4.  $\frac{\quad}{100}$ = ____%

Write as a percent.

5. $\frac{18}{100}$ = ____% **6.** $\frac{24}{100}$ = ____% **7.** $\frac{11}{100}$ = ____% **8.** $\frac{100}{100}$ = ____%

9. $\frac{46}{100}$ = ____% **10.** $\frac{37}{100}$ = ____% **11.** $\frac{1}{100}$ = ____% **12.** $\frac{3}{100}$ = ____%

Write as a fraction with a denominator of 100.

13. 87% ____ **14.** 25% ____ **15.** 2% ____ **16.** 17% ____

17. 33% ____ **18.** 56% ____ **19.** 66% ____ **20.** 1% ____

Write as a fraction in simplest form.

21. 25% ____ **22.** 8% ____ **23.** 20% ____ **24.** 60% ____

25. 10% ____ **26.** 15% ____ **27.** 75% ____ **28.** 17% ____

PROBLEM SOLVING

29. In a tile design of 100 tiles, 39 of the tiles are red. What percent of the tiles are red? ____________

30. A survey showed that 85 out of 100 people questioned were happy with the mayor of the town. What percent of the people questioned were happy with the mayor? ____________

31. In a group of 100 children, 48 had brown hair. What percent of the children had brown hair? ____________

Relating Percents to Decimals

Name ______________________

Date ______________________

Percent → Decimal	Decimal → Percent	
37% → 37. → 0.37	0.43 → 0.43 → 43%	Write 7 dimes as a percent of a dollar.
2% → 02. → 0.02	0.07 → 0.07 → 7%	7 × $.10 = $.70 0.70 = 70%
		So 7 dimes is 70% of a dollar.

Write as a decimal.

1. 43% ______ 2. 16% ______ 3. 10% ______ 4. 33% ______
5. 1% ______ 6. 72% ______ 7. 18% ______ 8. 68% ______
9. 14% ______ 10. 25% ______ 11. 52% ______ 12. 77% ______

Write as a percent.

13. 0.47 ______ 14. 0.05 ______ 15. 0.71 ______ 16. 0.99 ______
17. 0.08 ______ 18. 0.17 ______ 19. 0.54 ______ 20. 0.09 ______
21. 0.64 ______ 22. 0.59 ______ 23. 0.36 ______ 24. 0.85 ______

Write each as a percent of a dollar.

25. 8 dimes ______ 26. 7 nickels ______ 27. 4 pennies ______
28. 3 quarters ______ 29. 73 pennies ______ 30. 1 quarter ______
31. 5 dimes ______ 32. 3 nickels ______ 33. 18 pennies ______
34. 1 quarter, 7 nickels ______ 35. 3 dimes, 7 pennies ______
36. 11 nickels, 4 pennies ______ 37. 1 half dollar, 3 nickels ______

PROBLEM SOLVING

38. Ron spent 43¢. What percent of a dollar did he spend? ______________________

39. Luis added 0.38 liters of water to a solution he was mixing in chemistry class. What percent of a liter of water did he use? ______________________

40. Wynton sprinted 0.65 the length of a 100-meter long soccer field. What percent of the length of the field did he sprint? ______________________

 Use with Lesson 13-5, text pages 424–425.

Finding the Percent of a Number

Name ________________

Date ________________

45% of 80 = ?

Use Decimals

45% = 0.45

0.45 × 80 = 36

So 45% of 80 is 36.

Use Fractions

$45\% = \frac{45}{100} = \frac{9}{20}$

$\frac{9}{\cancel{20}_1} \times \cancel{80}^4 = 36$

Find the percent of the number by writing the percent as a decimal.

1. 45% of 40 ______
2. 15% of 60 ______
3. 60% of 40 ______
4. 20% of 90 ______
5. 7% of 300 ______
6. 8% of 350 ______

Find the percent of the number by writing the percent as a fraction.

7. 50% of 86 ______
8. 25% of 44 ______
9. 10% of 70 ______
10. 15% of 480 ______
11. 45% of 200 ______
12. 90% of 450 ______

Find the percent of the number.

13. 75% of 380 ______
14. 27% of 500 ______
15. 80% of 230 ______
16. 30% of 150 ______
17. 32% of 550 ______
18. 65% of 260 ______
19. 70% of 310 ______
20. 16% of 125 ______
21. 4% of 275 ______

Compare. Use <, =, or >.

22. 20% of 50 ______ 25% of 100
23. 40% of 60 ______ 60% of 40
24. 80% of 20 ______ 75% of 20
25. 20% of 20 ______ 50% of 40
26. 35% of 500 ______ 45% of 500
27. 60% of 60 ______ 90% of 40

PROBLEM SOLVING

28. Don had 280 baseball cards. He gave 15% of them to his sister. How many cards did Don give to his sister? ______________

29. There were 600 tickets sold for the dance recital. Trixie sold 5% of all tickets sold. How many tickets did Trixie sell? ______________

Using Percent

Name ______________________________

Date ______________________________

A sale on sweaters offers a discount of 15% off the regular price of $40.00.
How much is the discount?

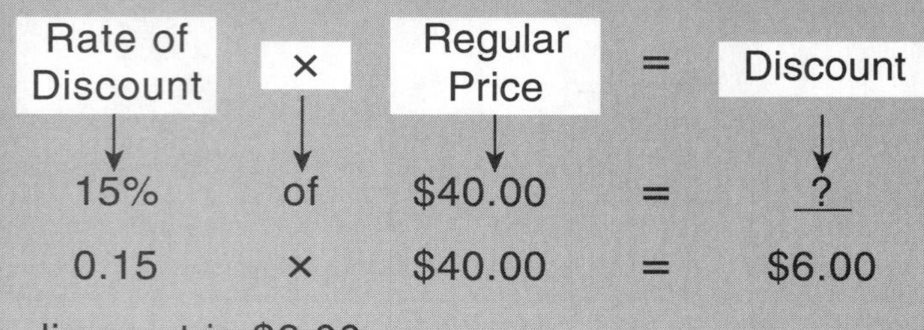

The discount is $6.00.

Complete the table. Use the circle graph at right.
Entertainment Plus rented 480 videotapes last week.

Weekly Rental

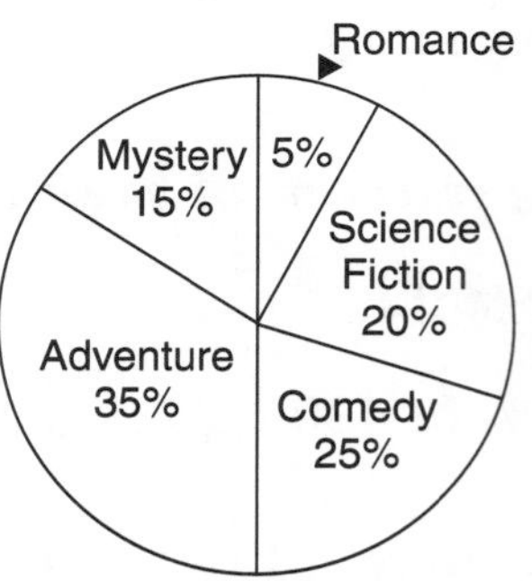

	Videotape	Percent	Number of Tapes Rented
1.	Adventure		
2.	Mystery		
3.	Romance		
4.	Comedy		
5.	Science Fiction		

Jaclyn earns $2000 a month.
How much does she spend for each.

Monthly Expenses

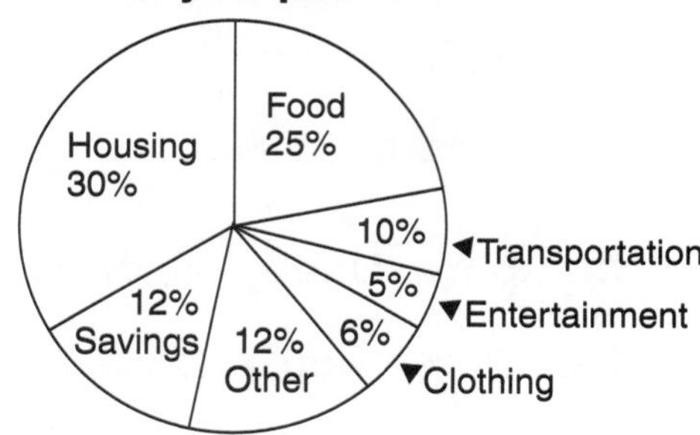

6. housing ________ **7.** food ________

8. savings ________ **9.** transportation ________

10. entertainment ________ **11.** clothing ________

Complete the table.

	Item	Regular Price	Rate of Discount	Discount
12.	sweater	$64	15%	
13.	jacket	$250	35%	
14.	sweat suit	$50	8%	
15.	jump suit	$125	20%	

PROBLEM SOLVING

16. Computers with a regular price of $1980 are offered at a 12% discount. What is the discount? ______________

17. Printers with a regular price of $320 are offered at a 7% discount. What is the discount? ______________

Problem-Solving Strategy: Combining Strategies

Name ______________________

Date ______________________

Palvik saves $1 the first week, $3 the second week, and $5 the third week. If he continues to save this way, how much will he have saved after 6 weeks?

First, *find a pattern* to tell how much he saves in the fourth, fifth, and sixth weeks. Each week, he saves $2 more than he did the preceding week.
Then, *use a table* to find the total amount he saved.

Palvik saved a total of $36.

Week	1	2	3	4	5	6
Amount	$1	$3	$5	$7	$9	$11
Total	$1	$4	$9	$16	$25	$36

Solve. Do your work on a separate sheet of paper.

1. Cathy bought a bag of wheat chips. Alan took 4 chips, and Paula took 3. She gave 4 to her mother and ate 3 herself. There were 6 chips left in the bag. What percent of the chips originally in the bag does Cathy have left?

   ____

2. Willis is making omelets for himself and 6 friends. He has 2 dozen eggs. He uses 2 eggs to make each omelet, and each person gets 1 omelet. How many eggs are left over?

   ____

3. Sharess has 100 rare coins. Twelve of the coins are silver dollars, 9 are silver quarters, 18 are silver dimes, 6 are buffalo nickels, and the rest are copper two-cent pieces. What percent of the coins are copper two-cent pieces?

   ____

4. Ramon uses 20 shells to make one necklace. Twenty-five percent of the shells are large shells and the rest are small shells. If Ramon wants to make 14 necklaces, how many large shells and how many small shells will he need?

   ____

5. Jane earns money delivering newspapers. She saves $2 the first week, $3 the second week, and $4 the third week. If she continues saving this way, how much money will she have saved by the end of the sixth week?

   ____

6. Donna's mother gives her money for a savings account. She gives $1 the first week, $5 the second week, and $9 the third week. If she continues saving this way, how many weeks will it take to save $45?

   ____

7. Emilio has 189 books to put onto 9 shelves. Some shelves hold 15 books and others hold two dozen. There are 60 shelves in one wing of the library. How many of each size shelf will Emilio use?

   ____

8. Four out of every 9 people surveyed said they would vote for Smith. If 80 people said they would vote for Smith, how many people were surveyed?

   ____

Problem Solving: Review of Strategies

Name ______________________

Date ______________________

Solve. Do your work on a separate sheet of paper.
Use the diagram at the right for problems 1–4.

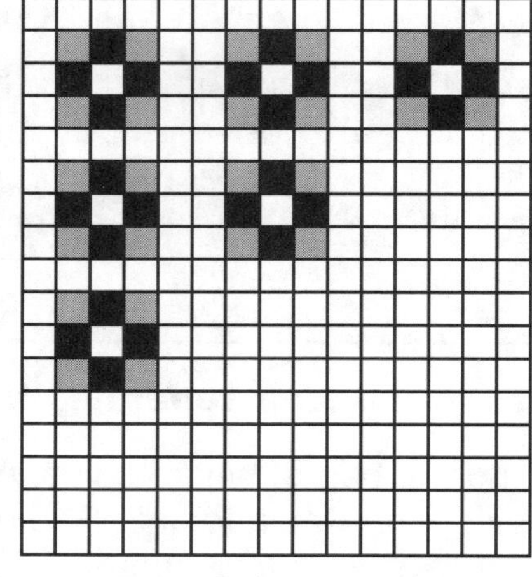

1. The Webber family is making a quilt for the craft show. The pattern they are using is begun for you in the diagram at the right. Shade the diagram to show how the quilt will look when it is finished.

2. What fractional part of the whole quilt is

   black? ______________________

   gray? ______________________

   white? ______________________

3. If each piece of material is a 5-inch square, what are the dimensions of the quilt?

   ____

4. If the Webber family decides to add a border of black squares around the quilt, how many squares will be needed to make the entire quilt? What fractional part of the whole quilt will be black?

   ____

5. A wildebeest can run at a speed of 50 mph. A white-tailed deer can run 60% as fast as a wildebeest, and a cheetah can run 140% as fast as a wildebeest. How many miles per hour faster than a deer does a cheetah run?

   ____

6. Alicia makes toy trucks to sell. Each truck requires 4 wooden wheels that she cuts from round dowels. Each wheel is $\frac{3}{8}$ inch thick. A dowel is 36 inches long. Alicia plans to make 60 trucks, how many dowels will she need?

   ____

7. Joanne reports that the average life expectancy of a gorilla is 20 years, although at least one lived for more than 39 years. The average life expectancy of a horse is also 20 years. However, at least one horse lived for 46 years. How much longer did the oldest horse live than the oldest gorilla?

   ____

8. Art and Loida work at a factory that is open 7 days a week. Art works for 5 days and then has 2 days off. Loida works for 3 days and then has 1 day off. Neither Art nor Loida works on Sunday, June 1, and both work on June 2. What other days in June do both Art and Loida have off?

   ____

Expressions: Addition and Subtraction

Name ______________________

Date ______________________

Numerical Expressions	Algebraic Expressions
$2 + 3$	$2a + 3x$
$10 - 5$	$10 - y$

Write an addition and a subtraction expression for each number.

1. 7 ______ ______

2. 32 ______ ______

3. 125 ______ ______

4. $\frac{7}{8}$ ______ ______

5. 945 ______ ______

6. $\frac{11}{12}$ ______ ______

7. 87 ______ ______

8. 401 ______ ______

9. 1200 ______ ______

10. $\frac{1}{3}$ ______ ______

Label each expression *numerical* or *algebraic.*

11. $2 + 3 + 8$ ______________

12. $9y - 4$ ______________

13. $d + 24$ ______________

14. $8 - 3 - 1$ ______________

15. $46 - 9$ ______________

16. $77 + 90w$ ______________

17. $x - y - z$ ______________

18. $11 + 63$ ______________

19. $4b + 4c$ ______________

20. $r - 10$ ______________

21. n ______________

22. $13 + 7$ ______________

Evaluate each expression.

23. $t - 0$, when $t = 8$ ________

24. $311 + b$, when $b = 145$ ________

25. $r - 4\frac{1}{2}$, when $r = 9$ ________

26. $p - 17$ when $p = 34$ ________

27. $1 + u$, when $u = 2000$ ________

28. $722 - c$, when $c = 11$ ________

29. $z - 99$, when $z = 100$ ________

30. $\ell + 80$, when $\ell = 6$ ________

31. $44 + f$, when $f = 55$ ________

32. $1900 + j$, when $j = 96$ ________

33. $g + h - i$, when $g = 2$, $h = 5$ and $i = 7$ ________

34. $a + b + c$, when $a = \frac{1}{2}$, $b = \frac{2}{3}$ and $c = \frac{1}{6}$ ________

35. $m - n + p$, when $m = 459$, $n = 32$, and $p = 199$ ________

Use with Lesson 14-1, text pages 440–441.

Expressions: Multiplication and Division

Name ______________________

Date ______________________

Numerical Expressions	Algebraic Expressions
12×2	$4 \times a$
$12 \cdot 2$	$4 \cdot a$
$12(2)$ or $(12)(2)$	$4(a)$ or $(4)(a)$
$15 \div 3$ or $\frac{15}{3}$	$4a$
	$y \div 5$ or $\frac{y}{5}$

Write a multiplication and a division expression for each number.

1. 12 ________ ________

2. 36 ________ ________

3. 5 ________ ________

4. $\frac{3}{4}$ ________ ________

5. 72 ________ ________

Name the operation shown in each expression.

6. $10n$ ________

7. $\frac{64}{a}$ ________

8. $b + 30$ ________

9. $d \cdot 5$ ________

10. $h - 15$ ________

Evaluate each expression.

11. $17 \times q$, when $q = 4$ ________

12. $150 \div k$, when $k = 25$ ________

13. $5r$, when $r = 75$ ________

14. $12 \cdot v$, when $v = 3$ ________

15. $\frac{n}{9}$, when $n = 45$ ________

16. $p \div 100$, when $p = 600$ ________

17. $(d)(22)$, when $d = 2$ ________

18. $19(m)$, when $m = 23$ ________

19. $343 \div b$, when $b = 7$ ________

20. $\frac{56}{x}$ when $x = 8$ ________

21. $10 \cdot e$, when $e = 13$ ________

22. $(9)(h)$, when $h = 11$ ________

Circle the letter of the correct answer. Which expression is equal to:

23. 450, when $r = 9$?

a. $50r$ **b.** $\frac{450}{r}$ **c.** $45 \cdot r$

24. 60, when $s = 5$?

a. $15 \cdot s$ **b.** $\frac{300}{s}$ **c.** $(s)(60)$

25. 36, when $t = 144$?

a. $36 \cdot t$ **b.** $\frac{t}{4}$ **c.** $36 \div t$

26. 573, when $u = 191$?

a. $573 \div u$ **b.** $3u$ **c.** $\frac{573}{u}$

Function Tables

Name ______________________

Date ______________________

Let m = 1 month. Let $\frac{m}{12}$ = number of years.

m	5	11	12	36	60
$\frac{m}{12}$	$\frac{5}{12}$	$\frac{11}{12}$	1	3	5

Complete each table.

1. Let s = weight of fruit in pounds. Let $s + 4$ = weight of fruit plus weight of box.

s	12	29	35	51	68
$s + 4$					

2. Let z = number of books. Let $\$7z$ = total cost.

z	1	3	5	7	9
$\$7z$					

3. Let b = perimeter of garden. Let $\frac{b}{4}$ = length of one side of garden.

b	16	40	52	64	100
$\frac{b}{4}$					

Write the rule for each table.

4.

a	?
3	12
5	20
8	32
12	48

____

5.

r	?
2	15
4	17
7	20
9	22

____

6.

f	?
75	15
40	8
15	3
5	1

____

Use with Lesson 14-3, text pages 444–445.

Addition Equations

Name ______________________

Date ______________________

Solve: $n + 16 = 30$ — Think: What number plus 16 = 30?

Try 12. $12 + 16 \stackrel{?}{=} 30$ — No, $28 < 30$

Try 13. $13 + 16 \stackrel{?}{=} 30$ — No, $29 < 30$

Try 14. $14 + 16 \stackrel{?}{=} 30$ — Yes, $30 = 30$

So $n = 14$.

Write the letter of the correct solution.

1. $r + 8 = 17$ ______
   a. $r = 5$ **b.** $r = 25$ **c.** $r = 9$

2. $45 = 12 + g$ ______
   a. $g = 24$ **b.** $g = 33$ **c.** $g = 23$

3. $a + 4 = 32$ ______
   a. $a = 28$ **b.** $a = 36$ **c.** $a = 4$

4. $15 + y = 75$ ______
   a. $y = 60$ **b.** $y = 90$ **c.** $y = 50$

5. $w = 8 + 19$ ______
   a. $w = 9$ **b.** $w = 11$ **c.** $w = 27$

6. $b + 20 = 25$ ______
   a. $b = 5$ **b.** $b = 25$ **c.** $b = 45$

Solve the equation.

7. $34 + x = 69$ ______ **8.** $e + 5 = 19$ ______ **9.** $n = 9 + 8$ ______

10. $c + 21 = 21$ ______ **11.** $43 = r + 7$ ______ **12.** $28 + 8 = f$ ______

13. $11 = 10 + y$ ______ **14.** $14 + w = 70$ ______ **15.** $a + 19 = 28$ ______

16. $b = 12 + 8$ ______ **17.** $p + 62 = 90$ ______ **18.** $76 = 14 + s$ ______

19. $\frac{4}{5} = t + \frac{1}{5}$ ______ **20.** $1\frac{1}{2} + d = 2$ ______ **21.** $j + \frac{1}{8} = \frac{1}{2}$ ______

Solve for the variable.

22. Perimeter = 56 dm

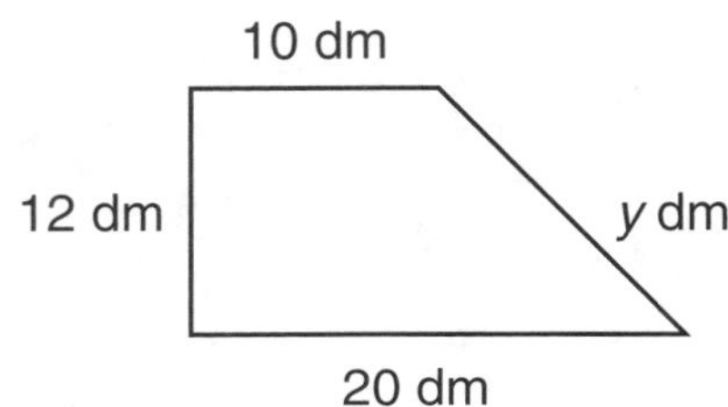

____

23. Perimeter = 19 yd

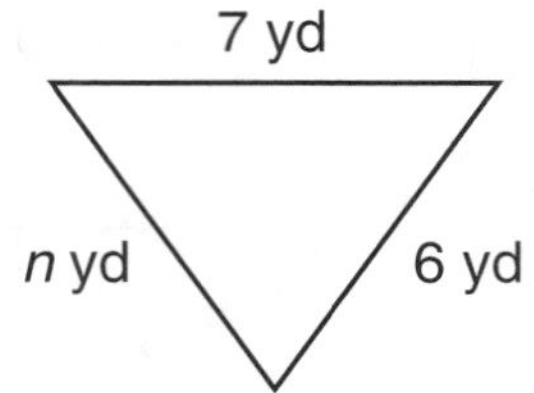

____

 Use with Lesson 14-4, text pages 446–447.

Multiplication Equations

Name ______________________

Date ______________________

Solve: $\frac{1}{5} \cdot y = 8$ Think: $\frac{1}{5}$ of what number = 8?

Try 30. $\frac{1}{5} \cdot 30 \stackrel{?}{=} 8$ No, $6 \neq 8$

Try 35. $\frac{1}{5} \cdot 35 \stackrel{?}{=} 8$ No, $7 \neq 8$

Try 40. $\frac{1}{5} \cdot 40 \stackrel{?}{=} 8$ Yes, $8 = 8$

So $y = 40$.

Write the letter of the correct solution.

1. $8w = 64$ ________

a. $w = 12$ **b.** $w = 8$ **c.** $w = 32$

2. $100 = 5(b)$ ________

a. $b = 10$ **b.** $b = 25$ **c.** $b = 20$

3. $12 \cdot r = 108$ ________

a. $r = 9$ **b.** $r = 108$ **c.** $r = 10$

4. $v \times 7 = 63$ ________

a. $v = 7$ **b.** $v = 56$ **c.** $v = 9$

5. $(z)(24) = 96$ ________

a. $z = 3$ **b.** $z = 4$ **c.** $z = 6$

6. $302 = 151(m)$ ________

a. $m = 2$ **b.** $m = 4$ **c.** $m = 151$

Solve the equation.

7. $12c = 144$ ________ **8.** $60 = 4(e)$ ________ **9.** $9 \cdot r = 117$ ________

10. $49 = 7t$ ________ **11.** $(u)(3) = 90$ ________ **12.** $69 = w \times 23$ ________

13. $n(29) = 0$ ________ **14.** $280 = 7g$ ________ **15.** $y \times 5 = 750$ ________

16. $10i = 110$ ________ **17.** $47(v) = 47$ ________ **18.** $12m = 600$ ________

19. $\frac{1}{2} \cdot 50 = h$ ________ **20.** $14 = \frac{1}{7} \times a$ ________ **21.** $\frac{1}{4} \cdot 24 = g$ ________

Solve for the variable.

22. Volume = 144 m^3

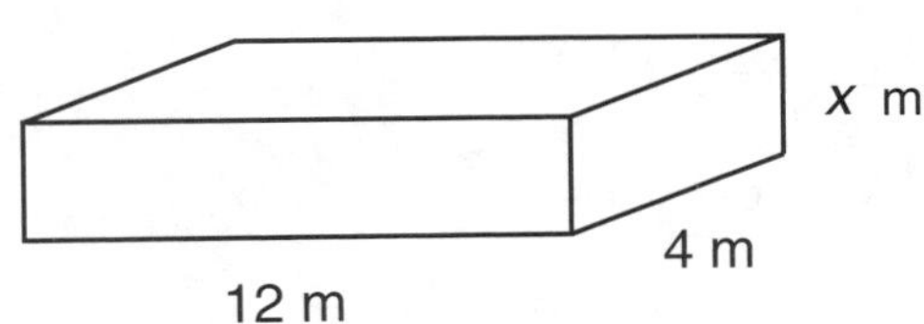

____

23. Area = 72 ft^2

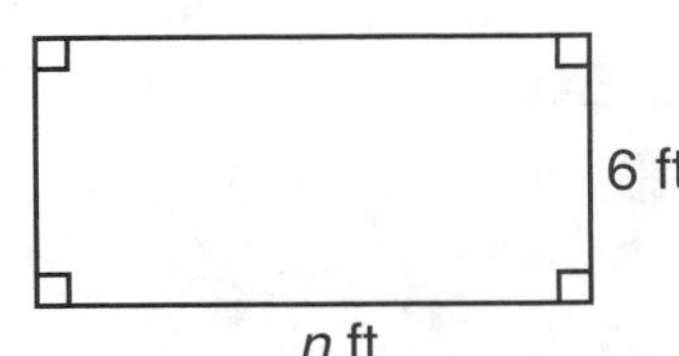

____

 Use with Lesson 14-5, text pages 448–449.

Fractions in Algebra

Name ____________________

Date ____________________

Equation	Property	Solution
$x + \frac{3}{7} = \frac{3}{7}$	Identity Property of Addition	$x = 0$
$\frac{1}{5} + \frac{3}{5} = \frac{3}{5} + x$	Commutative Property of Addition	$x = \frac{1}{5}$
$\frac{3}{4} \cdot x = \frac{3}{4}$	Identity Property of Multiplication	$x = 1$
$(4 \cdot \frac{1}{3}) \cdot x = 4 \cdot (\frac{1}{3} \cdot \frac{1}{4})$	Associative Property of Multiplication	$x = \frac{1}{4}$
$x \cdot \frac{4}{7} = 0$	Zero Property of Multiplication	$x = 0$

Solve for *m*. Use the property shown to help you.

Identity Property of Addition

1. $\frac{5}{6} + m = \frac{5}{6}$ ________
2. $m + 0 = \frac{3}{8}$ ________

Commutative Property of Addition

3. $m + \frac{1}{3} = \frac{1}{3} + \frac{4}{5}$ ________
4. $\frac{2}{3} + m = \frac{1}{4} + \frac{2}{3}$ ________

Zero Property of Multiplication

5. $m \times \frac{1}{7} = 0$ ________
6. $\frac{7}{11} \cdot m = 0$ ________

Identity Property of Multiplication

7. $1 \cdot m = \frac{5}{12}$ ________
8. $m \cdot \frac{15}{16} = \frac{15}{16}$ ________

Associative Property of Multiplication

9. $(5 \cdot \frac{1}{3}) \cdot m = 5 \cdot (\frac{1}{3} \cdot 3)$ ________
10. $(3 \cdot \frac{1}{9}) \cdot m = 3 \cdot (\frac{1}{9} \cdot \frac{1}{3})$ ________

Solve the equation. Use *properties* or *Guess and Test.*

11. $\frac{4}{7} \cdot p = \frac{4}{7}$ ________
12. $\frac{7}{8} = r + \frac{3}{8}$ ________
13. $(\frac{2}{3} + \frac{1}{4}) + 4 = \frac{2}{3} + (\frac{1}{4} + b)$ ________
14. $\frac{1}{3} = e + \frac{1}{12}$ ________
15. $\frac{1}{12} = \frac{1}{6} \cdot x$ ________
16. $\frac{7}{11} \cdot \frac{4}{5} = \frac{4}{5} \cdot w$ ________
17. $\frac{1}{6} \cdot (\frac{1}{3} \cdot m) = (\frac{1}{6} \cdot \frac{1}{3}) \cdot \frac{1}{5}$ ________
18. $a \cdot \frac{13}{19} = 0$ ________
19. $\frac{9}{13} = f + \frac{6}{13}$ ________
20. $\frac{2}{9} \cdot t = \frac{2}{9}$ ________
21. $c + \frac{5}{7} = \frac{5}{7} + \frac{3}{8}$ ________
22. $b + 0 = \frac{7}{10}$ ________
23. $\frac{5}{12} + u = \frac{5}{12}$ ________
24. $\frac{6}{7} \cdot j = \frac{3}{5} \cdot \frac{6}{7}$ ________
25. $\frac{3}{4} = d \cdot \frac{3}{4}$ ________
26. $d \cdot \frac{7}{8} = 0$ ________

 Use with Lesson 14-6, text pages 450–451.

Coordinate Geometry

Name ______________________

Date ______________________

Point A is located at ($^-3$, 2).
Point B is located at (1, 3).
Point C is located at ($^-1$, $^-2$).
Point D is located at (0, 3).

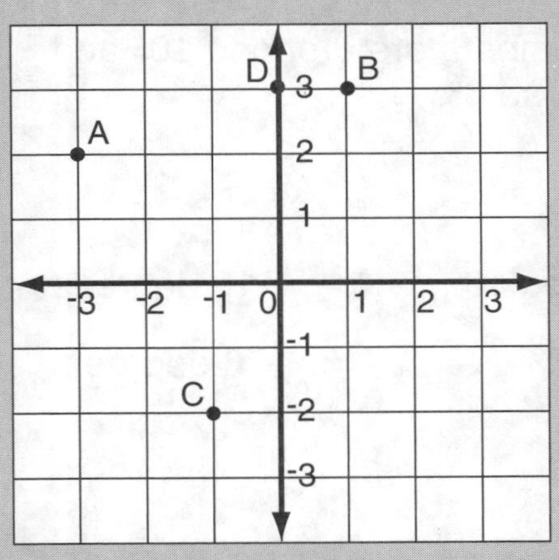

Use the grid at the right for exercises 1 – 12.
Name the point for each set of coordinates.

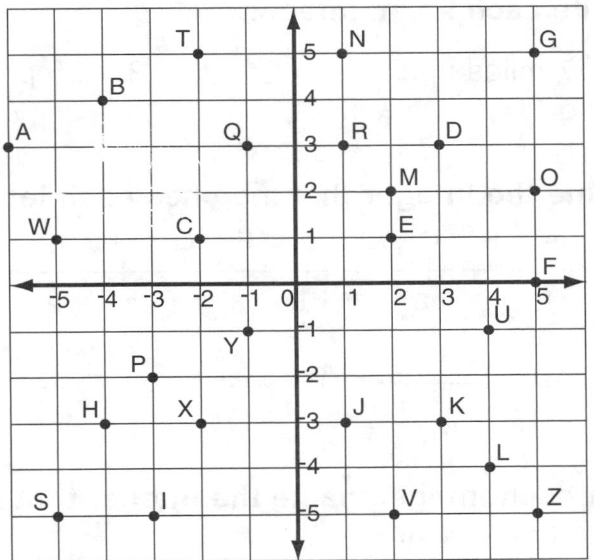

1. ($^-4$, 4) ______
2. (2, 1) ______
3. ($^-4$, $^-3$) ______
4. (3, $^-3$) ______
5. (1, 3) ______
6. (2, 2) ______
7. ($^-1$, 3) ______
8. (5, 0) ______
9. (4, $^-4$) ______
10. ($^-2$, 5) ______
11. ($^-5$, 1) ______
12. (2, $^-5$) ______

Complete the chart below.
Use the grid on the right.

	Point	Coordinates
13.		($^-4$, 2)
14.		($^-3$, 4)
15.		($^-1$, 4)
16.		(0, $^-1$)
17.		($^-3$, $^-3$)
18.		($^-4$, $^-1$)
19.		(3, 5)
20.		(5, 2)
21.		(1, 2)

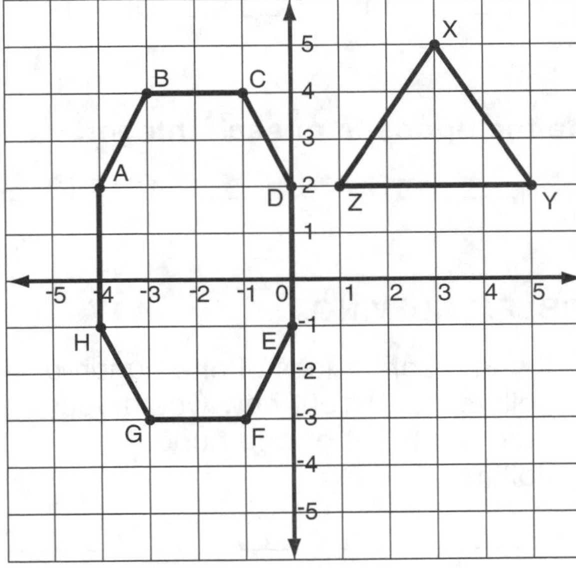

Introduction to Integers

Name ___________________________

Date ___________________________

The temperature dropped 10° and then rose 6°.

dropped 10° — $^{-}10$ degrees

rose 6° — $^{+}6$ degrees

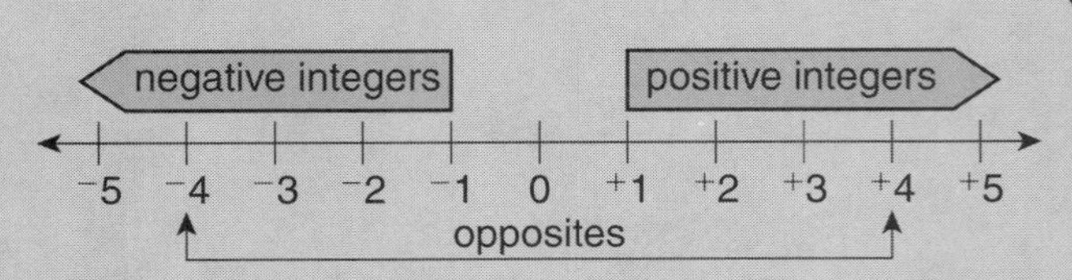

Positive (+) integers: *right* of 0; *greater than* 0

Negative (−) integers: *left* of 0; *less than* 0

Zero: neither + nor −.

Opposites: $^{+}4$ and $^{-}4$; 0 and 0; $^{+}1$ and $^{-}1$

Write each as an integer.

1. 7 miles uphill ________ **2.** \$5 profit ________ **3.** loss of 6 yards ________

Name the integer that matches each letter on the number line.

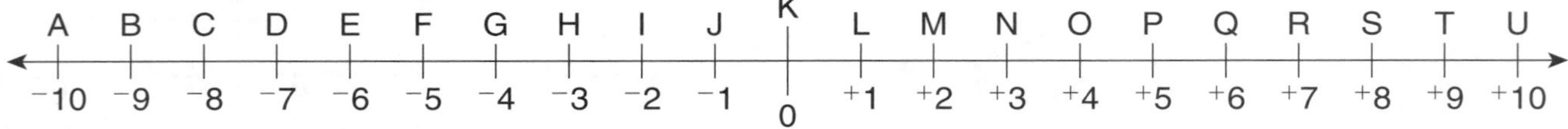

4. *K* ______ **5.** *A* ______ **6.** *P* ______ **7.** *S* ______ **8.** *M* ______ **9.** *E* ______

For each integer, name the integer that is just *before* and just *after* it on a number line.

10. $^{+}8$ ______ **11.** $^{-}7$ ______ **12.** $^{-}12$ ______ **13.** $^{+}2$ ______ **14.** 0 ______ **15.** $^{-}23$ ______

Write the opposite of each integer.

16. $^{+}1$ _____ **17.** $^{-}8$ _____ **18.** $^{+}13$ _____ **19.** 0 _____ **20.** $^{-}13$ _____ **21.** $^{+}92$ _____

PROBLEM SOLVING

22. If you record a loss of one hundred dollars as $^{-}$\$100, how would you record a profit of one hundred dollars?

____

23. On a map if you label the ground floor of an office building 0, how would you label the third level of the garage below the ground floor?

____

 Use with Lesson 14-8, text pages 454–455.

Compare and Order Integers

Name ______________________

Date ______________________

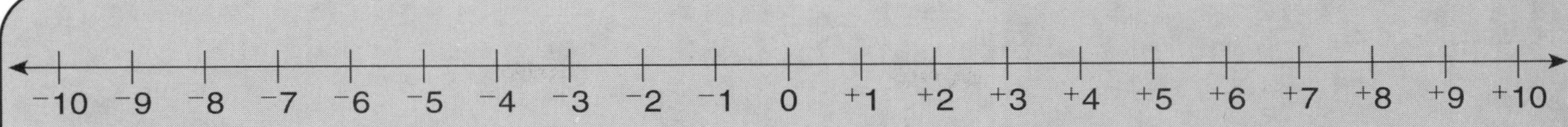

Compare Integers

Any integer is *greater than* an integer to its *left.*

$^{+}6 > {^{+}1}$ since $^{+}1$ is to the left of $^{+}6$.
Also $^{+}1 < {^{+}6}$.

$^{-}8 < {^{+}2}$ since $^{+}2$ is to the right of $^{-}8$.
Also $^{+}2 > {^{-}8}$.

Order integers

Least to greatest: Begin with integer farthest to the *left.*
$^{-}5$, $^{-}3$, 0

Greatest to least: begin with integer farthest to the *right.*
0, $^{-}3$, $^{-}5$

Choose the greater integer.

1. $^{+}1$, $^{+}8$ ______ **2.** $^{-}2$, $^{+}2$ ______ **3.** $^{-}5$, $^{-}4$ ______ **4.** $^{-}10$, $^{-}11$ ______

Compare. Write $<$ or $>$.

5. $^{-}6$ ______ $^{+}3$ **6.** $^{+}9$ ______ $^{-}1$ **7.** 0 ______ $^{-}3$ **8.** $^{-}10$ ______ $^{-}15$

Arrange in order from least to greatest.

9. $^{+}5$, 0, $^{-}7$ __________ **10.** $^{-}9$, $^{-}1$, $^{+}4$ __________ **11.** $^{-}10$, $^{+}2$, $^{-}13$ __________

Arrange in order from greatest to least.

12. $^{+}10$, $^{+}6$, $^{-}6$ __________ **13.** $^{-}8$, $^{-}12$, 0 __________ **14.** $^{+}6$, 0, $^{+}16$ __________

Write *always, sometime,* or *never* to make true statements.

15. A negative integer is __________ greater than a positive integer.

16. A negative integer is __________ greater than another negative integer.

17. Zero is __________ greater than a negative integer.

PROBLEM SOLVING

18. In her checkbook register Brenda records deposits as $^{+}\$100$ and withdrawals as $^{-}\$100$. For the following entries:
$^{+}\$100$, $^{-}\$18$, $^{+}\$50$, $^{-}\$55$, and $^{-}\$45$,
which withdrawal is between \$40 and \$50? __________

Adding Integers with Like Signs

Name ______________________________

Date ______________________________

To model adding integers on the number line:
Add: $^-4 + {}^-1 = \underline{?}$

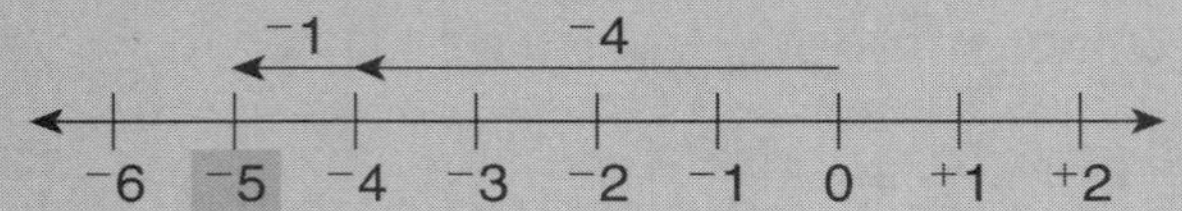

- Start at 0.
- Move *left* for negative integers.
- Move *right* for positive integers.
- Zero means *no* move. $^-4 + {}^-1 = {}^-5$

To add integers with *like* signs:
Add: $^-4 + {}^-1 = \underline{?}$

- Add the integers. $4 + 1 = 5$
- Use the sign of the addends. $^-4 + {}^-1 = {}^-5$

Write an addition sentence for each number line.

1.

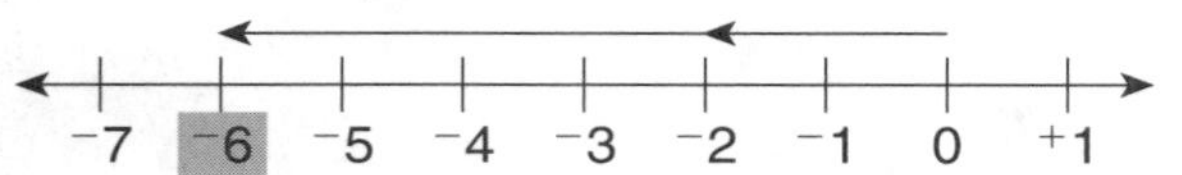

____

2.

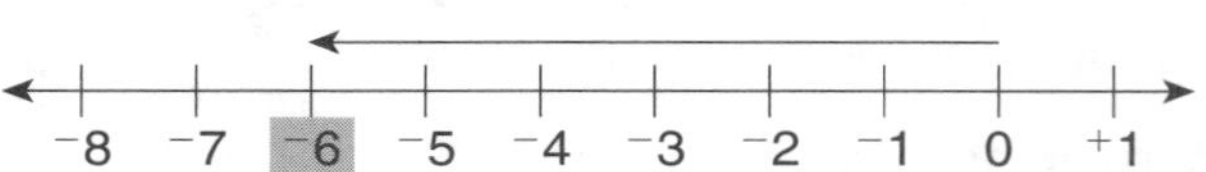

____

Add. Use a number line to help you.

3. $^-6 + {}^-3$ ______ **4.** $^+9 + {}^+6$ ______ **5.** $0 + {}^-3$ ______ **6.** $^-10 + {}^-15$ ______

Complete each table.

7.

n = integer	$n + {}^+3$
$^+6$	
0	
$^+4$	

8.

n = integer	$n + {}^-3$
$^-5$	
$^-3$	
$^-7$	

Find the sum.

9. $(^+3 + {}^+3) + {}^+6$

________ + $^+6$ = ________

10. $^-4 + (^-6 + {}^-1)$

$^-4$ + ________ = ________

11. $^-4 + (^-3 + {}^-8) =$ ______ **12.** $(^-7 + {}^-9) + 0 =$ ______ **13.** $(^-8 + 0) + {}^-9 =$ ______

PROBLEM SOLVING Write the answer in words and as an integer.

14. A golfer shot three strokes above par on the first hole and then one stroke above par on the next hole. How many strokes above par is her score on the two holes? ______________________

 Use with Lesson 14-10, text pages 458–459.

Adding Integers with Unlike Signs

Name ______________________

Date ______________________

To model adding integers on the number line:

Add: $^{+}4 + {}^{-}1 = \underline{?}$

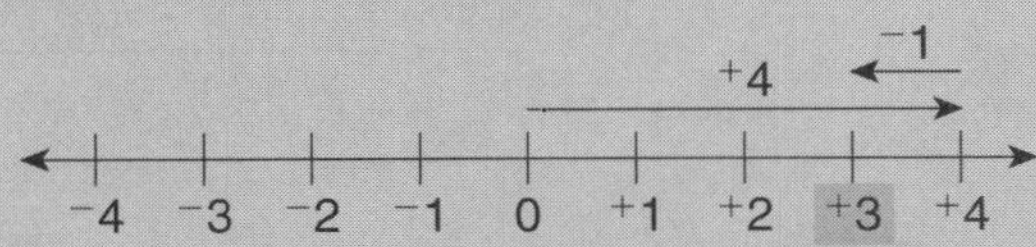

- Start at 0.
- Move *left* for negative integers.
- Move *right* for positive integers. $^{+}4 + {}^{-}1 = {}^{+}3$

To add integers with *unlike* signs:

Add: $^{+}4 + {}^{-}1 = \underline{?}$

- Find the difference. (Drop the signs and subtract the numbers.) $4 - 1 = 3$
- Use the sign of the addend farther from zero. $^{+}4 + {}^{-}1 = {}^{+}3$

$^{+}4$ is farther from zero. The sign is +.

Complete the addition sentence for each number line.

1.

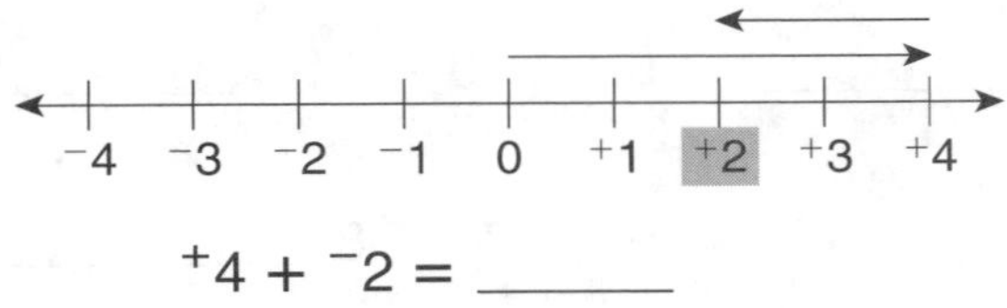

$^{+}4 + {}^{-}2 =$ ______

2.

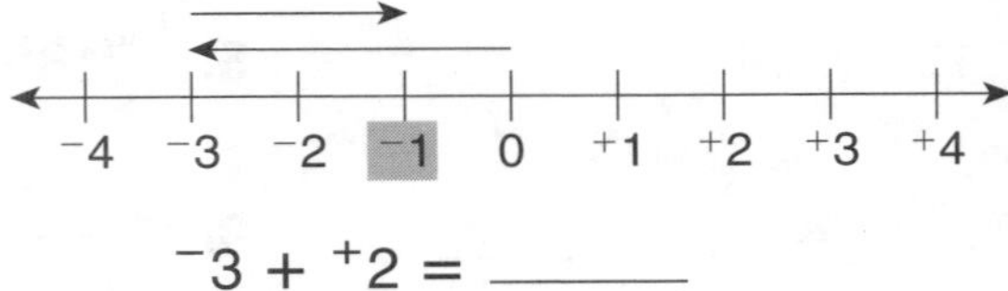

$^{-}3 + {}^{+}2 =$ ______

Write addition sentence for each number line.

3.

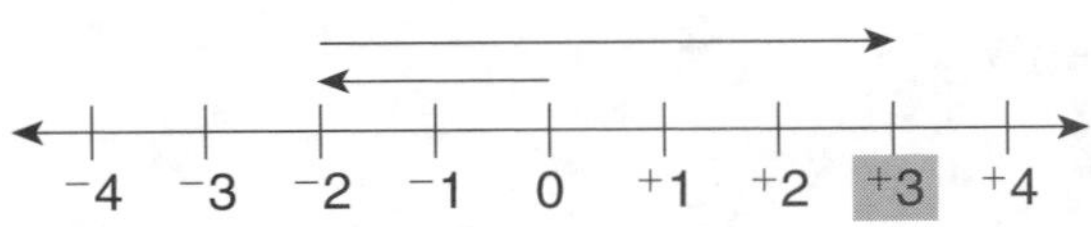

____

4.

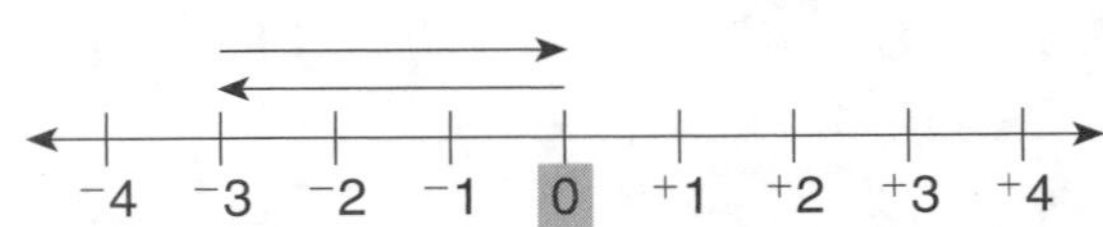

____

Find the sum. Use a number line to help you.

5. $^{-}6 + {}^{+}3$ ______
6. $^{-}9 + {}^{+}6$ ______
7. $^{-}10 + {}^{+}3$ ______
8. $^{-}10 + {}^{+}15$ ______
9. $^{+}11 + {}^{-}5$ ______
10. $^{+}7 + {}^{-}7$ ______
11. $^{+}9 + {}^{-}6$ ______
12. $^{+}10 + {}^{-}4$ ______

PROBLEM SOLVING Write each answer in words and as an integer.

13. A newborn baby lost 7 ounces in the hospital and gained 5 ounces his first week at home. What is his net gain or loss?

____

14. Flora climbed 45 feet up the side of a cliff. Then she slipped 10 feet down. Where is she located now?

____

Subtracting Integers

Name ______________________________

Date ______________________________

Subtracting an integer is the same as *adding the opposite* of that integer.

Subtract: $^{-}3 - {}^{-}5 = ?$

$^{-}3 - {}^{-}5 = {}^{-}3 + {}^{+}5$

$^{-}3 - {}^{-}5 = {}^{+}2$

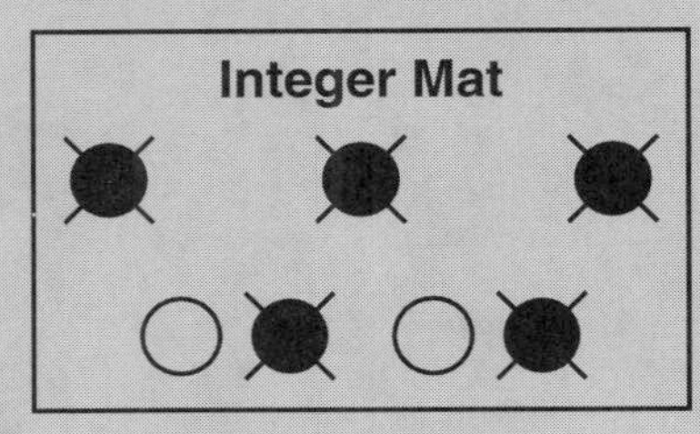

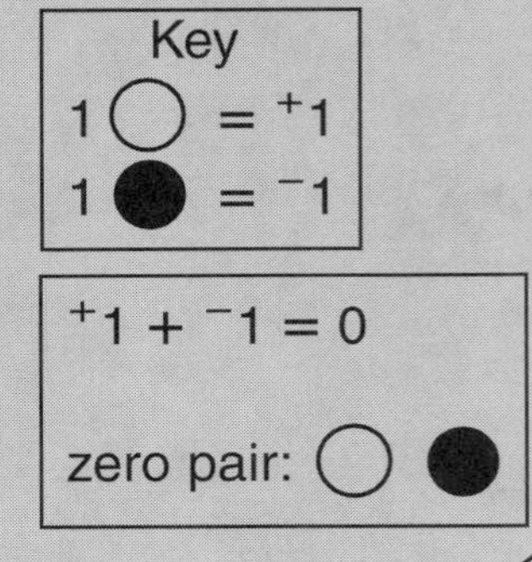

Rewrite each as an addition expression. Then add.

1. $^{-}7 - {}^{+}2$ __________ **2.** $^{+}6 - {}^{-}2$ __________ **3.** $^{-}9 - {}^{-}7$ __________

4. $^{-}12 - {}^{-}14$ __________ **5.** $^{+}18 - {}^{-}5$ __________ **6.** $^{-}10 - {}^{-}10$ __________

7. $^{-}5 - {}^{+}4$ __________ **8.** $^{+}8 - {}^{-}11$ __________ **9.** $^{+}7 - {}^{+}8$ __________

Subtract. Use counters to help you.

10. $^{+}6 - {}^{-}8$ ______ **11.** $^{-}9 - {}^{-}2$ ______ **12.** $^{-}6 - {}^{-}11$ ______

13. $^{-}4 - {}^{-}6$ ______ **14.** $^{+}6 - {}^{+}2$ ______ **15.** $^{-}1 - {}^{-}5$ ______

Circle the letter of the correct answer.

16. $^{-}12 - {}^{-}1$ **a.** $^{-}13$ **b.** $^{-}11$ **c.** $^{+}13$ **d.** $^{+}11$

17. $0 - {}^{-}13$ **a.** $^{-}13$ **b.** $^{+}13$ **c.** 0 **d.** $^{-}1$

18. $^{-}7 - {}^{-}7$ **a.** 0 **b.** $^{-}14$ **c.** $^{+}14$ **d.** $^{+}1$

PROBLEM SOLVING

19. On a winter day, the temperature dropped from $^{-}3$°C to $^{-}11$°C. Find the change in temperature.

____

20. Kay lives 4 blocks north of school. Joe lives 7 blocks south of school. What is the distance between Kay's house and Joe's?

____

Multiplying Integers

Name ______________________

Date ______________________

The product of two integers:

is *positive* if they have the *same* sign.	is *negative* if they have *different* signs	is *zero* if one or both is *zero*.
${}^{+}5 \times {}^{+}2 = {}^{+}10$	${}^{-}5 \times {}^{+}2 = {}^{-}10$	$0 \times {}^{-}2 = 0$
${}^{-}4 \times {}^{-}3 = {}^{+}12$	${}^{+}4 \times {}^{-}3 = {}^{-}12$	$0 \times {}^{+}2 = 0$
		$0 \times 0 = 0$

Use the rules above to find each product.

1. ${}^{-}6 \times {}^{+}3$ ______
2. ${}^{-}9 \times 0$ ______
3. ${}^{+}10 \times {}^{-}3$ ______
4. ${}^{-}10 \times {}^{-}5$ ______
5. ${}^{-}3 \times {}^{+}3$ ______
6. ${}^{+}3 \times {}^{+}3$ ______
7. ${}^{-}3 \times {}^{-}3$ ______
8. $0 \times {}^{+}3$ ______
9. ${}^{-}9 \times {}^{-}3$ ______
10. ${}^{+}6 \times {}^{+}3$ ______
11. ${}^{-}4 \times {}^{+}5$ ______
12. ${}^{+}8 \times {}^{+}3$ ______
13. ${}^{+}7 \times {}^{-}6$ ______
14. ${}^{-}8 \times {}^{+}4$ ______
15. ${}^{+}9 \times {}^{+}3$ ______
16. ${}^{-}5 \times {}^{-}9$ ______

Let *p* = positive integer and *n* = negative integer. Choose the correct answer (a. positive, b. negative, and c. zero) to complete each statement. Explain each answer.

13. $n \times p =$ ______

____

14. $(n \times p) \times p =$ ______

____

15. $(0 \times n) \times p =$ ______

____

16. $n \times 0 =$ ______

____

17. $(p \times p) \times p =$ ______

____

18. $n \times (n \times n) =$ ______

____

PROBLEM SOLVING

19. A healthcare stock gains 2 points each day for five days. What is the net gain over the five days?

____

20. A pipe was leaking water at a rate of 5 gallons an hour. What was the net loss of water over a four-hour period?

____

Use with Lesson 14-13, text pages 464–465.

Dividing Integers

Name ______________________

Date ______________________

Multiplication Sentence	Related Division Sentences
$^-5 \times {}^+3 = {}^-15$	$^-15 \div {}^+3 = {}^-5$ $^-15 \div {}^-5 = {}^+3$
$^+4 \times {}^-5 = {}^-20$	$^-20 \div {}^-5 = {}^+4$ $^-20 \div {}^+4 = {}^-5$
$^-5 \times {}^-6 = {}^+30$	$^+30 \div {}^-6 = {}^-5$ $^+30 \div {}^-5 = {}^-6$

The quotient of two integers:

- is *positive* if they have the *same* sign
  $^+10 \div {}^+2 = {}^+5$
  $^-9 \div {}^-3 = {}^+3$
- is *negative* if they have *different* signs.
  $^-10 \div {}^+2 = {}^-5$
  $^+9 \div {}^-3 = {}^-3$

Complete.

1. $^-9 \times {}^+3 = {}^-27$
   $^-27 \div {}^+3 =$ ______
   $^-27 \div {}^-9 =$ ______

2. $^-8 \times {}^-6 = {}^+48$
   $^+48 \div {}^-6 =$ ______
   $^+48 \div {}^-8 =$ ______

3. $^+4 \times {}^+6 = {}^+24$
   $^+24 \div {}^+6 =$ ______
   $^+24 \div {}^+4 =$ ______

Write two related division sentences.

4. $^-7 \times {}^+2 = {}^-14$

5. $^+6 \times {}^-2 = {}^-12$

6. $^-9 \times {}^-7 = {}^+63$

7. $^+8 \times {}^+9 = {}^+72$

8. $^-7 \times {}^-6 = {}^+42$

9. $^+5 \times {}^-9 = {}^-45$

Divide.

10. $^+64 \div {}^-8$ ______
11. $^-90 \div {}^-2$ ______
12. $^+24 \div {}^-8$ ______
13. $^-56 \div {}^-4$ ______
14. $^-42 \div {}^+6$ ______
15. $^-12 \div {}^-1$ ______
16. $0 \div {}^-11$ ______
17. $^-66 \div {}^-11$ ______
18. $^-72 \div {}^-8$ ______
19. $^+81 \div {}^+9$ ______
20. $^-39 \div {}^-3$ ______
21. $^-24 \div {}^-4$ ______

PROBLEM SOLVING

22. A scuba diver dives to a depth of 150 feet in 25 minutes. What is the average rate of the dive per minute, written as an integer?

23. A water pump pumps out a basement filled with 45,000 gallons of water in 9 hours. What is the average amount per hour, written as an integer?

 Use with Lesson 14-14, text pages 466–467.

Functions and Coordinate Graphs

Name____________________________________

Date____________________________________

Graph the function $y = x + {}^{-}2$ on a coordinate grid.

- Make a function table.
- Graph each ordered pair. Connect the points.

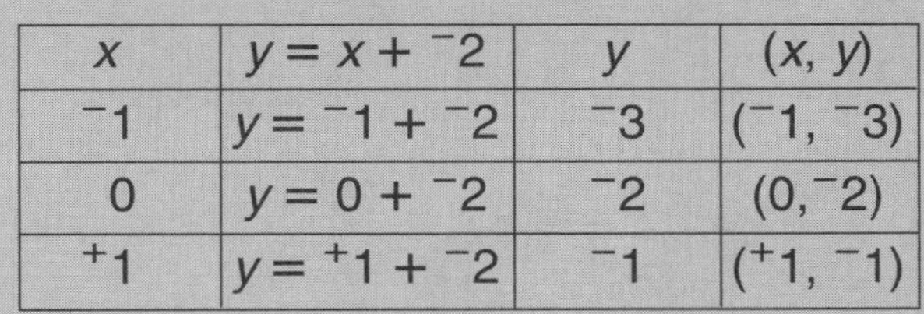

x	$y = x + {}^{-}2$	y	(x, y)
$^{-}1$	$y = {}^{-}1 + {}^{-}2$	$^{-}3$	$({}^{-}1, {}^{-}3)$
0	$y = 0 + {}^{-}2$	$^{-}2$	$(0, {}^{-}2)$
$^{+}1$	$y = {}^{+}1 + {}^{-}2$	$^{-}1$	$({}^{+}1, {}^{-}1)$

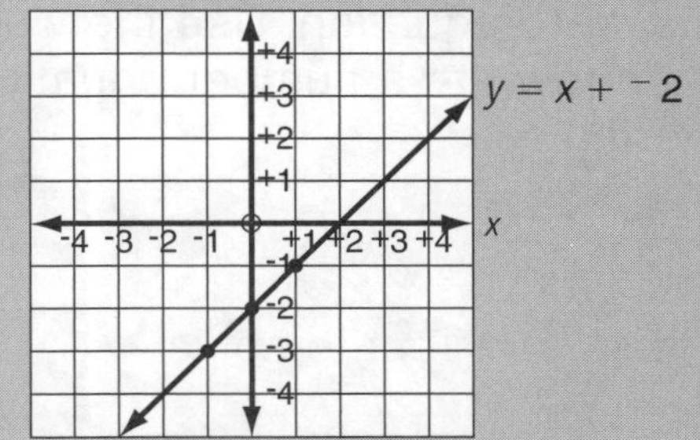

1. Complete the function table. Then graph on the coordinate grid.

x	$y = x + {}^{-}1$	y	(x, y)
$^{-}1$			
0			
$^{+}1$			
$^{+}2$			

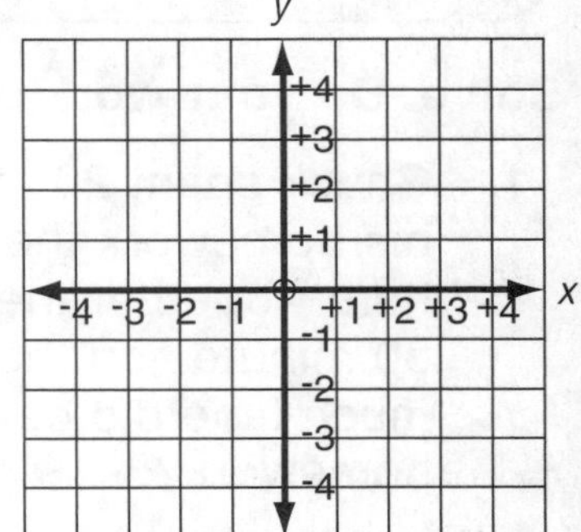

Use the given graph of $y = x + {}^{-}4$

2. When $x = 0$, what is the value of y? ________

3. When $x = {}^{+}2$, what is the value of y? ________

4. For what value of x is $y = {}^{-}1$? ________

5. For what value of x is $y = 0$? ________

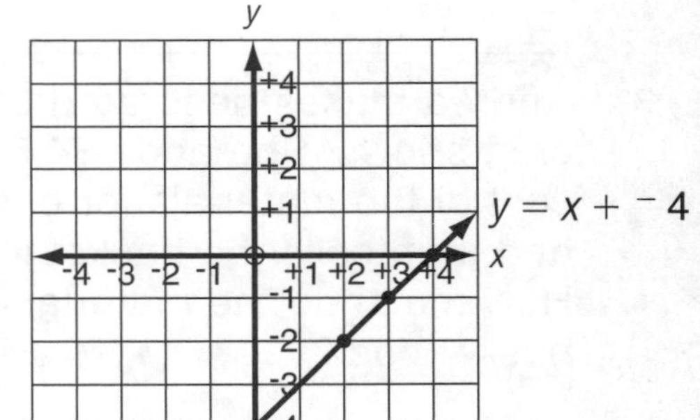

PROBLEM SOLVING

6. A meteorologist discovered that a storm is following a path on her map made by the equation $y = x + {}^{-}3$. Will the storm pass through the point $({}^{+}2, 0)$? Make a function table. Then graph on a coordinate grid to answer.

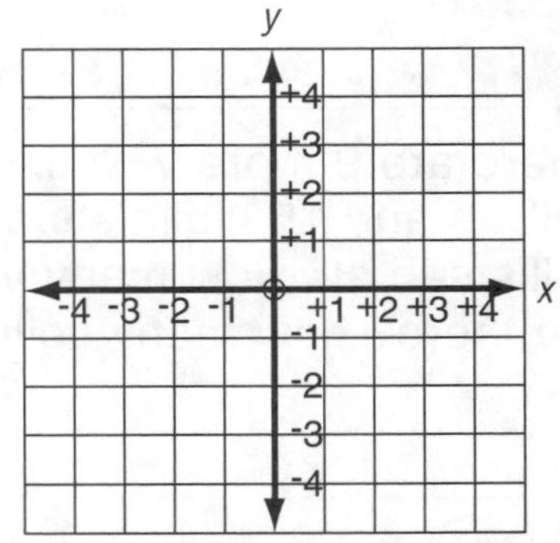

____

Use with Lesson 14-15, text pages 468–469.

Problem-Solving Strategy: Write an Equation

Name ______________________

Date ______________________

Lionel is 3 years older than his sister Luella. If the sum of their ages is 27, how old is each?

Write a word equation. Use the word equation to write a number equation with variables.

Let y represent Luella's age.
Then $y + 3$ represents Lionel's age.

Luella's age plus Lionel's age is 27

$y + (y + 3) = 27$

Use Guess and Test to solve.

Try 10: $10 + 13 = 23 \rightarrow 23 \neq 27$
Try 15: $15 + 18 = 33 \rightarrow 33 \neq 27$
Try 12: $12 + 15 = 27 \rightarrow 27 = 27$

Luella is 12, and Lionel is 15.

Solve. Do your work on a separate sheet of paper.

1. Chuck spent 20 minutes longer doing his homework than Stacy took doing hers. Together they spent 2 hours and 30 minutes on their homework. How much time did each spend on homework?

____

2. Hank, Carlos, and Mindy formed a country music trio. They practiced $1\frac{1}{2}$ hours on Friday and $1\frac{3}{4}$ longer than that on Saturday. How long in all did they practice on Friday and Saturday?

____

3. Andrea makes decorated baskets at craft shows. She adds $4.50 to the cost of the materials for each basket and then sells the basket for $9.95. How much do the materials for each basket cost?

____

4. Ginny and Gerald each have one coin. The value of Ginny's coin is 15¢ more than the value of Gerald's coin. The total value of both coins is 35¢. What coin does each person have?

____

5. There are 18 people in the Computer Club. There are 2 more girls than there are boys. How many girls and how many boys are in the club?

____

6. If you double Kari's age and add 2, you will get Ted's age. The sum of their ages is 20. How old is each person?

____

7. There are 9 more violinists than cellists and 15 violinists and cellists in all. How many violinists are there? how many cellists?

____

8. Juwon has 250 baseball cards of National League players and $\frac{1}{2}$ that number of American League players. How many baseball cards does he have in all?

____

 Use with Lesson 14-16, text pages 470–471.